1   2007

# Agriculture, Nematology and the Society of Nematologists

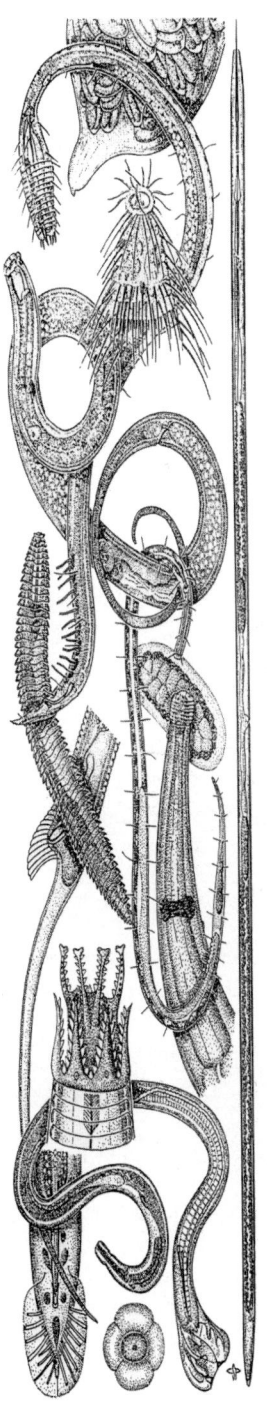

# Agriculture, Nematology and the Society of Nematologists

D. P. Schmitt
&
J. M. P. Schmitt

 SCHMITT & ASSOCIATES OF MARCELINE
MARCELINE, MISSOURI

Agriculture, Nematology and the Society of Nematologists
D. P. Schmitt and J. M. P. Schmitt

Published by Schmitt & Associates of Marceline
14844 Highway 5, Marceline, Missouri, 64658, U.S.A.

Distributed by The SON Business Office, P. O. Box 311, Marceline, Missouri, 64658, U.S.A.

All rights reserved. No part of this book may be reproduced or transmitted in any form or by any means, electronic or mechanical, or by any information storage or retrieval without written permission from the respective author or authors, except for the inclusion of brief quotations in a review.

Copyright © 2007 under the title of Agriculture, Nematology and the Society of Nematologists by the Society of Nematologists.

Library of Congress Control Number: 2007902760

ISBN: 9780975322918

Printed by Walsworth Publishing Company, Marceline, Missouri, U.S.A.
Printed in the United States of America.

Cover is a montague of diverse nematodes drawn by Charlie Papp and S. A. Sher. The artwork was used as the official Journal of Nematology cover from 1969 to 2000.

This book is printed on acid-free paper.

*Book design by Julia Marie P. Schmitt.*

*Dedicated to all nematologists,
past, present and yet to come.*

# Contents

| | | |
|---|---|---|
| Preface | | ix |
| 1 | An Introduction | 1 |
| 2 | The Nematological Explosion | 7 |
| 3 | Call for Nematologists | 13 |
| 4 | Creation and Birth of the Society of Nematologists | 23 |
| 5 | The Golden Age of Nematology | 37 |
| 6 | The Evolution of Nematology | 51 |
| 7 | Dawning of a New Generation | 65 |
| 8 | The New Millennium | 73 |
| 9 | The Nematological Horizon | 83 |
| References | | 87 |
| Appendix I - Society of Nematologists Presidents | | 95 |
| Appendix II - Society of Nematologists Officers | | 99 |
| Appendix III - Journal of Nematology Editor-in-Chiefs | | 101 |
| Appendix IV - Nematology Newsletter Editors | | 103 |
| Appendix V - Annual Meetings | | 105 |
| Appendix VI - Society of Nematologists Constitution and Bylaws | | 107 |
| Appendix VII - Nathan.A. Cobb Nematology Foundation Bylaws and Constitution | | 123 |
| Appendix VIII - Nematological Quotes | | 127 |
| Index | | 133 |

# Preface

*An institution cannot be understood without first understanding the conditions that led to its creation.*
                                                            -*J. K. Galbraith*

For the past five decades, laboratories, greenhouses and field experiments have been abandoned for the duration of one week enabling a committed group of individuals, dedicated to the study of nematology, to scamper around a conference center in some remote corner of the world carrying SON satchels and exchanging information. Each year, a privileged few are called to active duty a day earlier than the others in order to sit in executive board meetings, dedicating their time and talent to brightening the future of the Society of Nematologists. At one such session, a decision was made to further the Society by looking back at its history. Members at the Second Executive Board Meetings at Ithaca, New York at the Forty-second Annual Meeting in 2003 stated, "the Society should consider a history of nematology since the treatment in *Vistas on Nematology* was very terse." By some twist of fate, the responsibility of this endeavor fell on our shoulders.

The big question that surfaced immediately was, "how does one write a history of a unique group of scientists that in the scheme of things, belong to a relatively young organization, working in a field of research unknown to most?" In our opinion, there was only one way to approach a history of SON, as an examination of this organization from the perspective of those who have been instrumental in its formation, promulgation and continuation. Countless interviews were conducted as well as hundreds of hours dedicated to the examination of the Society of Nematologists Archives, which contains vital information about the Society from its formative years to the present day.

Once research into the history of the Society of Nematologists began, it was discovered that the decisions which shaped the formation and direction of the Society were affected by the events that took place in nematology and agriculture during specific decades from the 1950's until the present. This book has given us the opportunity to examine trends and key events which occurred from the Society's conception to the present in an attempt to capture the trials, joys and accomplishments of those who dreamt a dream that came to fruition in the embodiment of the Society of Nematologists.

This work would not have been made possible without the assistance of many individuals. We would like to thank all past presidents of the Society especially James Baldwin, Kenneth Barker, Burton Endo, Virginia Ferris, Stephen Lewis, Michael McClure, Terry Niblack, Andy Nyczepir, Robert Riggs, Rodrigo Rodriguez-Kabana, Gerald Santo, Joseph Sasser, James Starr, Seymour Van Gundy, Diana Wall and John Webster. The presidents listed have provided valuable information for this book. We also would like to express our gratitude to Robert Aycock, David Bird, Rob Brown, Ed Bouska, Ed Caswell-Chen, Van Cotter, Larry Duncan, Howard Ferris, Ian King, James Mueller, Don Norton, Charles Opperman, Charles Tarjan and Melissa Yoder for providing insight which furthered our understanding of social, economic and agricultural issues that impacted SON. Also, we would like to thank the Iowa State University Special Collections personnel for their assistance and cooperation. Special thanks to Anna Marie P. Schmitt for her assistance with research and editing, and Cecilia Marie P. Schmitt, for her consultation of the graphic design and editing. All materials gathered for this book have been placed in the SON Archives at Iowa State University, Ames, Iowa.

*As any prophet knows, to understand the future, one must study the past. History puts the present into perspective and the future into logic, as far as the future can be predicted.*

-President W.B. Mountain
SON Presidential Farewell

# 1 An Introduction

Nematodes have been relentless companions of human beings since the beginning of time. Nathan A. Cobb, Father of American Nematology, observed that "if all the matter in the universe except the nematodes were swept away, our world would still be dimly recognizable; we would find its mountains, hills, valleys, rivers, lakes and oceans represented by a film of nematodes."

Great populations of nematodes have been built up over time. W. H. Hart stated, "[human beings] are often said to be their own worst enemy. He [or she] converts land to agricultural use, thus upsetting a natural balance and creating conditions which often favor the development of plant-parasitic organisms and suppress other organisms which might control them. In addition, [individuals] carry plants, plant parts and various commodities associated with plants from all corners of the world, [and] in so doing, does a most efficient job of disseminating plant pests and diseases."

During the early history of civilization, individuals obtained food by methods of hunting and gathering. Eventually, human beings provided for themselves by agricultural means. From this point until the twentieth century, the world community was fundamentally an agricultural society, with generation after generation sustaining themselves by the cultivation of their land, producing varieties of modern day wheat, barley, rye and corn. The majority of the adult working population was engaged in farming until the 1890's. In the 1880's, for example, 75% of all adults worked on farms.

The first shift from an agricultural based society to an urban centered one occurred in Europe. Largely due to the Industrial Revolution in the 1800's, Europe began to transform from an agricultural centered society to a town centered society engaged in manufacturing. While Europe was transforming itself, the majority of the population of the United States was still living in rural communities. The American population did not begin to experience a shift from an agrarian society to a commercial society until after the Civil War (1861-1865).

At the dawn of the twentieth century, the percentage of people engaged in farming had decreased to 40%. The Economic Research Service reported that "at the beginning of the 20th century, rural America was the center of American life. [Farms] were home to most of the [American] population and was the source of food and fiber for the Nation's sustenance and commerce. And most of the nations population was involved in producing that food and fiber. The typical rural community in 1900 consisted of a small town or village with numerous small farms within a few miles. Most people lived their lives and fulfilled most of their needs, economic and otherwise, within this community. They had little contact with areas beyond the community." Ed Bouska, an Iowa farmer born in 1911, commented that "if I die in bed some night, I will have died 10 feet from where I was born. I have been here [farming] all my life."

As a result of continuous planting without the use of crop rotation, farmers began to detect a substantial decrease in crop yields. Many scientists who were interested in solving agricultural problems during the early to middle 1900's were individuals who grew up on farms and witnessed the decrease in yields. Farming was a beneficial experience for scientists who could reference knowledge gained from their practical experiences in order to readily understand plant production problems. In this context, G. Thorne commented, "it is especially important that, whenever possible, workers should be assigned to problems in those sections of the country where they have grown up and with whose general agricultural picture they are already acquainted. A farm background is invaluable; no amount of technical training can take its place."

For centuries, curious scientists associated the decreases in crop yields to damage done by nematodes. In 1743, J. Needham was the first scientist in Europe to report the occurrence of the nematode commonly known as the wheat seed gall nematode. Thirty years later, Roffredi was the first to publish that nematodes caused disease in plants. However, it was not until the nineteenth century, when workers in the sugar cyst beet industry recognized the serious losses caused by the sugar beet nematode, that the economic significance of plant-parasitic nematodes was realized. It was due to this loss that nematode research received a high priority in the latter part of the 19th century.

For the next hundred years, continuous contributions were made in the field of plant nematology especially in the area of nematode taxonomy by European scientists such as K. Rudolphi, J. Kuhn, C. von Siebold, H. C. Bastian, O. Butschli, J. G. de Man, L. Oerley, H. Micoletzky and I. N. Filipjev. There were other contributions that were made but were not recorded. This omission in the literature was unfortunate as articulated by K. R. Barker, retired nematologist, North Carolina State University. He stated, "in the early days of nematology, there were many able people who made contributions locally which are not reflected in publications. Nematology benefited by their work but opportunities to increase the horizons of nematology were lost."

In the early agricultural period of the United States, plant-parasitic nematodes were partly responsible for the rapid loss in productivity along the Atlantic Coast. By 1832, it was estimated that in the state of Virginia, all the crops were worth less than the agricultural exports of the state ninety years before. George Washington, the First President of the United States, described the land as "originally very good, but use and misuse have made the [land] quiet otherwise." Although the agricultural problem was recognized, little attention was devoted to the development of nematode control measures during the first quarter of the twentieth century.

**Figure 1-1.** Nathan A. Cobb (1859-1932), Father of American Nematology.

Due to the research of Nathan A. Cobb, nematological advances occurred in the United States during the early 1900's. Through his research, he described 1000 nematode species, and developed equipment and apparati that enhanced the science. He also trained the first generation of American nematologists such as E. M. Buhrer, J. R. Christie, G. Steiner, A. L. Taylor and B. G. Chitwood. It was N. A. Cobb and his student G. Steiner who proposed the theory that nematodes could be the primary cause of plant disease.

According to G. Thorne, a colleague of Cobb, research "indicated that nematodes, perhaps even more than soil depletion and erosion, were the cause of land abandonment on an enormous scale. Fields were cleared, tobacco grown for four to five years when the crop usually failed. Then corn was grown until it too failed and the land was allowed to return to brush and forest and new fields cleared. This process continued until there was no longer excess land to be cleared and yields of tobacco and corn dropped to a small fraction of what they once were."

Influenced by Cobb, nematological emphasis was placed on zoological aspects, morphological characterizations, classification of the phylum and ultimately taxonomy and systematics. Only a small number of plant pathologists and plant breeders conducted research on diseases caused by plant-parasitic nematodes or on the development of resistant varieties prior to 1935.

During this time, those dedicated to researching nematode problems began corresponding with each other. Research notes, textbooks, microscope slides and other items were exchanged between peers forming nematological networks. In the Southeastern United States, tobacco scientists joined together to inform scientific and agricultural societies that nematodes were causing problems. At the Third National Plant Nematode Conference held in Birmingham, Alabama in 1940, a permanent plant nematode council was unanimously approved with the sole purpose of promoting research on problems caused by plant-parasitic nematodes and facilitating mutual assistance among researchers.

Despite these advances, the agricultural problem of crop losses due to nematodes was still not widely acknowledged up through the first half of the 1940's. H. H. Hume, Dean of the School of Agriculture (1938-1943), University of Florida, described the limited awareness of nematodes and their agricultural impact by stating, "if there were no nematodes in the South [of the United States] and they would suddenly appear in their present numbers, they would be seen as the pestilence they are. Funds for their control would be supplied quickly and in large amounts. Such was the situation which allowed complete eradication of the Mediterranean Fruit Fly. But the nematode problem arouses no particular interest. Nematodes are always working havoc, taking their toll on crops, sometimes causing complete destruction. We blind ourselves by accepting them as a matter of cause. The problem is here; we live with it; but there is no solution. There must be a general awakening all along the line to the magnitude of the situation."

The "awakening" of which H. H. Hume described was just around the corner. A sense of urgency to control nematodes would grip the country following World War II. J. N. Sasser commented, "I do not know any science that developed as rapidly and as thoroughly as nematology."

*All men can seek the truth and, as scientists...advance in their knowledge of nematodes, we, too, advance and gain knowledge with worldwide acquisition of data.*
*-President J. N. Sasser*
*SON Presidential Address*

# 2 The Nematological Explosion

After World War II, the world experienced a population boom never before seen. In the 1950's and 1960's, the need arose to produce enough food to sustain the ever growing population. The World Food and Agricultural Organization (FAO) placed major emphasis on sustainable agriculture. Anything that hindered agricultural production drew great attention.

One issue that surfaced in the 1940's and 1950's was the great damage and crop loss caused by plant-parasitic nematodes. Scientists and farmers alike recognized nematodes as a major problem that needed immediate attention. In the South, cotton, tobacco and vegetable crops were being devastated by root-knot nematodes. R. D. Riggs, retired nematologist, University of Arkansas, stated, "in certain areas they could hardly grow cotton any more. Soybean problems came later. It added to the situation, not only from the stand point of root-knot nematode but the eventual occurrence of soybean cyst nematode. Root-knot was a serious problem on tomatoes in areas like Florida and tomato growing areas around the country." J. M. Good observed that during World War II, "in the Southwestern Georgia counties, as much as 30% of the suitable peanut land was devoted to this crop. Under these conditions, there was little opportunity for crop rotation, thereby producing favorable conditions for root-knot nematode repro-

**Figure 2-1.** The World Food and Agricultural Organization (FAO) International Nematology Course and Symposia in Rothamsted, United Kingdom, September 3-14, 1951.

duction. In infested areas, peanut yield losses ranged from 20% to 95%." R. Aycock, Professor Emeritus and former Head of the Plant Pathology Department, North Carolina State University, commented, "root-knot especially was a problem in the South because of the climate: warm temperatures, sandy soils, etc. When people came back from World War II, students began to look into these problems. In North Carolina, there were problems on peaches; orchards would 'run out.' Tobacco had a terrific problem. And to some extent problems occurred on peanuts. People were very conscious of that...I think that had to be the basis of the great stimulus in the South. We needed control or management of these terrific problems. On tobacco, there was a lot of pressure to work on resistant varieties."

The growing devastation caused by nematodes was not limited to the southern states. D. Norton, retired nematologist, Iowa State University, recalled, "I think a lot of it was just finding out that nematodes were a problem in the northern states. When I came to Iowa State, I was told that the problems we had were not like the South. We found out more and more that there were a lot of problems. They did not know anything about the lesion nematode. We found out we had problems in Iowa on corn. One thing here in Iowa, a corn and soybean state, the losses seem small but cover a greater area than you find in a citrus or vegetable crop. It takes nematologists to go out and find them."

Although nematode problems in the United States were not enough to change farming practices, awareness was heightened in 1941 when the golden nematode, which had devastated the European and South American potato crops, was found on Long Island, New York by B. G. Chitwood. Fear that the US potato crop would be wiped out brought nematode control into the spotlight. Concern continued to increase when the tobacco cyst nematode (1951) and soybean cyst nematode (1954) were discovered in the USA. V. R. Ferris, nematologist, Purdue University, recalled, "there were a lot of people running around the country saying nematodes were going to kill us. There was a hidden enemy that no one knew about because it was microscopic. In the Midwest; in Missouri, Indiana, Illinois, Ohio; we all started to demonstrate damage. I cannot tell you how many years John [Ferris] and I tried to convince people that nematodes were reducing yields in corn. The entomologists stated, 'no, those are our organisms causing it.' Plant pathologists claimed that it was their organisms. It was hard to sort out what was causing the losses. It was the advent of the soybean cyst nematode which changed awareness of nematodes in the Midwest. It was the most harmful pest to soybean throughout that whole area and you could demonstrate the damage. Money began to be put into soybean cyst nematode research because the nematode was causing yield reductions."

Recognition of the damage caused by nematodes in the 1940's and 1950's in conjunction with the discovery of certain nematodes was tied directly to the economic impact on agricultural production. It was this economic impact that ultimately generated the awareness of nematodes. As E. L. Ayers so wittily stated, "the disclosure may not be enough to frighten a grower into listening to the facts on spreading decline or the burrowing nematode. But change the figures into cold cash and interest generally blossoms and grows."

The science of nematology received a major catalyst with the discovery and introduction of nematicides. In 1943, W. Carter reported that a dichloropropene-dichloropropane mixture from the Shell Oil Company was an effective nematicide. This nematicide was trade named D-D and placed on the market by Shell Chemical Company in 1943. In 1945, J. R. Christie reported that ethylene dibromide (EDB), a chemical used as an insecticide as early as 1925, was also effective for controlling root-knot nematode. This nematicide, marketed by Dow Chemical Company, provided farmers with a low cost and effective tool to control nematodes. In the 1950's, a variety of nematicides flooded the market. 1,2-dibromo-3-chloropropane (DBCP) was developed and sold as Nemagon Soil Fumigant by Shell Chemical Company and as Fumazone by Dow Chemical Company. DBCP was also sold under a wide variety of synonyms by several chemical companies. Other chemicals introduced during this era were Vapam and VC-13.

Nematicides not only provided a way to control nematodes but they also helped scientists demonstrate the impact of nematodes on crops and shined a spot light on nematology. As V. R. Ferris explained, "all of the early work on animal and plant-parasitic forms began as taxonomic and life history studies by curious biologists. The tremendous impact of plant-parasitic nematodes on crop production was not even realized until after the discovery and widespread use of the first soil fumigants." J. M. Webster, retired nematologist, Simon Fraser University, commented that "the development of chemical nematicides has been credited with being the savior of the

Hawaiian pineapple, East African tobacco industries and possibly the citrus industry of Florida in the 1950's. The eventual widespread use of nematicides on a wide range of food and fiber crops caused people to realize the enormous amount of crop damage and loss of yield that the insidious plant-parasitic nematodes were incurring. Undoubtedly, it can be said that D-D and EDB saved whole industries and significantly increased crop yields worldwide while putting nematology on the map."

With the introduction of nematicides, farmers needed to be trained in how to apply the chemicals. J. T. Sturgis, an extension worker, stated, "soil fumigation, being completely new, was a difficult practice to develop among the growers. First, we had to induce the growers to work with us on a field demonstration plot, with D-D being applied without cost and, in most cases, the application being made by us. I might add, such men as A. L. Taylor, now head of the USDA Section of Nematology, and Shell's own E. F. Feichtmeir, were among those who helped in this pioneer work. The results of the first year's work were more gratifying and effective control of the root-knot nematode was satisfactory." R. Aycock observed that "after [World War II], farmers were taught how to apply the nematicides with a gravity flow system pulled by a mule. It was a great thing. You could see marvelous differences." S. Van Gundy, Professor Emeritus of Nematology, University of California-Riverside commented that "you had an instant control. A farmer went out there and fumigated and instantly saw [the results]. That was dollars in his pocket. In those days, growers put it on as insurance. It was cheap insurance for the farmer. He saw it work so he put it on again the next year and the next year." By 1957, an estimated 500,000 acres were fumigated, resulting in a 400% increase in yields.

The use of pesticides and nematicides caused a great structural change in agriculture. T. J. Sheets stated, "greater acreages evolved to specialized, highly mechanized and chemical-intensive production systems. These systems dramatically increased yields and land use while decreasing labor requirements." At the same time that chemical nematicides were being discovered and fumigation equipment being developed, plant breeders started developing resistant varieties of crops such as tomato, soybean and tobacco. K. R. Barker stated, "this gave additional tools to demonstrate the effects of nematodes on crops. In parallel, resistance was developed to other pathogens-some of the fungi and bacteria-by a number of groups, which showed that nematodes caused direct damage but could also circumvent resistance to the pathogens by altering the structure, physiology and biochemistry of the hosts such as tomato or tobacco. Other growths in nematology included the development of damage functions, recognition of the importance of nematodes in disease complexes and virus spread, extensive descriptions of the structure and anatomy as well as function and behavior of nematodes, their physiology, cytogenetics and genetic diversity and systematics, descriptions and morphological aspects."

In 1951, Christie and Perry were the first to prove that *Trichodorus* spp. were plant parasites. The proof that this ectoparasitic nematode was a pathogen marked a major breakthrough in nematology. In 1950, B. G. Chitwood and M. B. Chitwood added greatly to nematode systematics with the publication of their work entitled *An Introduction to Nematology.*

With this new awareness of nematodes and the rapidly increasing technology of nematode control, a need for qualified individuals to work on the nematode problem became quickly evident.

*It is almost a Law of Plant Pathology that the number of pathogens of a given host is directly proportional to the number of professional workers assigned to that host.*

-A. A. Forest<br/>Shell Nematology Workshop

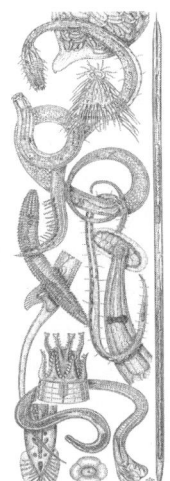

# 3 Call for Nematologists

Although nematological research was already being conducted prior to the 1950's, there was a great need for a group of men and women dedicated solely to the science of nematology. Two different groups of future nematologists were targeted: students (undergraduate and graduate) and plant pathologists who were already studying nematodes. According to R. D. Riggs, "there were nematologists around prior to that time but most of the ones I am familiar with were members of plant pathology departments who happened to be working on nematode problems amongst other things they were doing. In certain states, nematode problems were serious enough that they might have a nematologist, particularly in terms of root-knot nematode or the citrus nematode."

Initially, plant pathologists were called upon to work on these problems. In 1946, the Research and Marketing Act was passed by the US Congress which provided funding for research based on problems regional in scope. Four nematology projects were created and named the Southern Regional Project, the Northeastern Regional Project, the North Central Regional Project and the Western Regional Project. In order to train students and cross train practicing scientists in the specific area of nematology, a federal-state program, which was referred to as the Regional Research Project or Committee, facilitated research education in nematology during the 1950's.

In the South, agricultural experiment station directors recognized the pressing need for trained personnel to research the relationship between nematodes and plant diseases. They recommended that Regional Research Funds be allocated to institutions in positions to develop graduate teaching programs in nematology. The initial annual budget in 1953 for this project was $15,000. It increased to $135,000 by 1957.

In order to technically train research workers in the field of nematology, the Southern Regional Project, which was commonly referred to as the S-19 Technical Committee, annually held four workshops that last one to two weeks. E. Cairns, Chairman of the S-19 Technical Committee, stated that "these workshops were aimed at broadening the scope of the training and knowledge in phytonematology of participants in the project and of students majoring in this subject. Recognizing that the workers and teachers in this country are, perhaps much 'from the same mold', efforts were made to have someone come from abroad as a guest lecturer." D. Norton commented, "I was at Texas A&M as a plant pathologist. Word came down that they were having this short course in Raleigh, North Carolina in 1954 and they wanted somebody from each state. This was the first S-19 Southern Regional Committee Workshop. I was interested in root-knot so my boss said go ahead and attend. It was a two week session. Eldon Cairns, Joe Sasser and Al Taylor were instructors. Many people at that session went on to become members of the Society of Nematologists." J. N. Sasser stated, "at those meetings, we decided that if we could, we needed to hold a graduate course in nematology to help train people that were not nematologists but were excellent plant pathologists with an interest in nematology." D. Norton further explained that "many of the people present were mainly plant pathologists who would be converted into nematologists. After the workshop, I came back and started up a course in plant nematology. The other people did too."

Many scientists also attended short workshops sponsored by Shell Chemical Company during the 1950's. These workshops focused on the practical aspects of nematology and contributed greatly to the nematology training of graduate students and practicing plant pathologists working on nematological problems. At the meetings, almost every crop affected by nematodes in the United States was discussed. For example, at the Shell Nematology Workshop at Yuma, Arizona, the crops touched upon ranged from citrus to vegetables such as tomatoes, beans, garlic, sugar beets, squash, carrots and cucumbers; while the workshop held in Columbia, South Carolina focused on nematodes on lawn grass, forest trees, peanuts, tobacco and peaches. The solutions proposed at these meetings revolved heavily around soil fumigants and nematicides. S. Van Gundy credited these workshops with "really getting nematology rolling in California. This was a real driving force of momentum in the West." J. T. Sturges stated at the Shell Nematology Workshop in Toledo, Ohio that "in 1953, Shell [Oil] Company made available to us, experimentally, such quantities of OS 1897 [the experimental nematicide that was later marketed as DBCP] that we could apply on the field scale experiments."

These workshops began in 1953 and continued until 1959 when an explosive demand for nematologists made it necessary to initiate plant nematology undergraduate and graduate courses in universities as well as short courses in the US and Europe. Prior to World War II, most nematologists received their training from G. Thorne or a cooperative teaching program between the USDA and the University of Maryland. Plant Pathologists, entomologists and agricultural scientists also had been training themselves to work in the field of nematology. After the War, land-grant college enrollment

16   Agriculture, Nematology and SON

**Figure 3-1.** S-19 Nematology Workshop at North Carolina State University, Raleigh, North Carolina in 1954.

**Figure 3-2.** S-19 Nematology Workshop at North Carolina State University in Raleigh, North Carolina. From left to right: W. Hare, University of Mississippi; D. Steiner, USDA, Beltsville; L. Miller, Virginia Tech; J. Sasser, North Carolina State University; C. Wilson, Virginia Tech; and E. Cairns, Auburn University.

**Figure 3-3.** J. N. Sasser at Nematology Workshop at North Carolina State University, Raleigh, North Carolina in 1954.

**Figure 3-4.** S-19 Nematology Workshop Group photograph at North Carolina State University in Raleigh, North Carolina in 1954.

## 18  Agriculture, Nematology and SON

**Figure 3-5.** Phytonematology Seminar at the University of Tennessee, Knoxville, Tennessee, July 1-5, 1957.

increased substantially as veterans enrolled under the GI bill. During the 1940's, 584,000 students enrolled in agricultural courses compared to 21,000 in the 1920's.

In 1948, the first formal nematology course was offered at the University of California at Berkeley by M. W. Allen. In the 1950's and 1960's, G. Steiner and W. F. Jeffers taught nematology courses at the University of Maryland, B. G. Chitwood offered courses at Catholic University, Washington D. C., G. Weber started the first nematology class at the University of Florida, W. Mai offered classes at Cornell University and J. N. Sasser and H. Hirschmann taught the first year long series of two nematology courses at North Carolina State University. Also, at Cambridge University, F. G. W. Jones offered England's first short course in nematology while Imperial College at Silwood Park Field Station began the first educational program in nematology. As a result of these courses, the number of nematology teachers increased greatly and

Call for Nematologists 19

**Figure 3-6.** Shell Nematology Workshop in 1958.

**Figures 3-7.** A. L. Taylor presenting at the Shell Nematology Workshop in 1958.

by 1959, 13 institutions offered formal nematology courses. However, progress was limited. As J. R. Christie, in a letter to A. C. Tarjan, wrote, "none of the departments grant Ph.D. degrees in nematology. There are several departments where it is possible for a student to get a Ph.D. degree and write a thesis on some subject pertaining to nematodes."

Many of the advances that took place at the university level in the field of nematology were made possible through money that had accumulated during World War II. The funds were distributed through the federal and state governments making it very easy to receive funds. J. N. Sasser stated, "when I went to N.C. State, I did not know what a grant was. I just knew they hired me. They paid me. They paid for my travel. They paid for my graduate student stipends. I never dreamed of a grant. You were hired and the state said you had "X" number of dollars and this is your program. I knew they paid my salary. If I got a graduate student, they paid his stipend. If I wanted to go to a meeting, they paid my way. I had a technician. That was just the way it was." Money was also provided to universities by chemical companies to support nematology research. D. Norton stated, "when I was at Texas A&M, money just seemed to come. I did get some grants from some chemical companies, Dow Chemical and others, to do some more practical work with nematicides. In Iowa, it was easy to get grants from these chemical companies. It was also used to do some fundamental work."

Funds allocated to agricultural projects that supported the Experiment Stations came from the federal government and were based upon the number of individual farms in a state. The more a state's economy was based on agriculture, the more important the impact of nematodes on crop production became in that state. For example, R. D. Riggs commented, "anything that affected the agriculture of the state, affected the whole state. If they could demonstrate that there were enough prob-

Call for Nematologists 21

Figure 3-8. Quentin Holdeman, first editor of the Nematology Newsletter (1956-1958).

lems being created by a particular pest, then it was not difficult to get the legislature in any state to set up funds to support the management of that problem. In Arkansas, when the soybean cyst nematode was found, it did not take anytime for the legislature in Arkansas to create a position for a nematologist. The potential for damage to the soybean industry in the state at the time was so great that they did not hesitate."

With funds pouring into research and education, the number of individuals working in nematology increased and communications between these scientists became more frequent. During the early 1950's, individuals working with nematodes shared nematode specimens mounted on microscope slides, discussed characteristics, taxonomy and new species of nematodes. They exchanged papers and books about nematodes from Europe, South America, North America, Asia and Africa. Nematologists began meeting at annual meetings of the American Phytopathologial Society (APS) as well as at the different region technical committee sessions.

The members of the S-19 Technical Committee on plant-parasitic nematodes recognized the need to foster better communications between those working and studying in the field of nematology. In an effort to bridge this communication gap, Quentin Holdeman, plant pathologist, Pee Dee

Experiment Station at Florence, South Carolina, suggested at the Third Annual S-19 Technical Committee meeting that there was a need for a newsletter dedicated to nematology and volunteered to be the editor. The idea was accepted by the committee and Holdeman, without a budget, started the Nematology Newsletter in 1955. He was given permission to publish it under the auspices of the S-19 Technical Committee. In the first issue of the Nematology Newsletter, Holdeman wrote that the newsletter was being "distributed to members of the S-19 Committee and to the USDA section of nematology. It is hoped that it will provide a means of acquainting nematologists with each other's needs and problems." Even though this newsletter lacked the status of being a publication of an official organization, it was the primary vehicle of interaction between nematologists during the 1950's.

*A very few years ago there was hardly a man in the United States who could or would screen the soil and sample roots to determine the degree of infestation and foretell with great accuracy whether or not soil should be treated. Today there are many trained scientific minds working in nematology, not to mention the many entomologists, plant pathologists and laymen who have eagerly acquired a working knowledge of these hidden enemies of agriculture.*
        -L. F. Stayner
        Shell Nematology Workshop

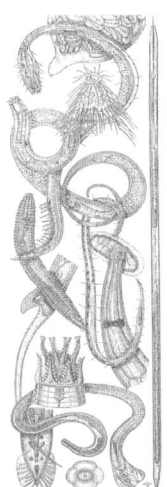

# 4 Creation and Birth of the Society of Nematologists

As the awareness of nematode-incited-problems increased during the 1940's and 1950's, the need to establish a permanent forum for the exchange of nematological information became apparent. By the late 1950's, funding from government and commercial sources for nematology programs experienced phenomenal growth. This funding supported the initiation of university courses on the subject and an increase in nematology symposia organized for the purpose of bringing together nematologists from Europe and from other countries around the world.

  Despite the rift between previously warring nations, during World War II, scientists were coming together to solve their common crop production problems. J. M. Webster stated that European nematologists "came together in a post-war environment which was still trying to mend the hurt of what had happened in the previous decade. Seven to eight years [earlier], they all had been at war with each other. They were talking because there were great constraints on food production. They knew they needed better food production. Sugar beet and potato staples were a major concern for the Europeans. Nematodes were part of that problem." In an effort to combat the ever growing nematode problem, nematologists joined together for the First International Nematology

Symposium which took place in 1951 in Harpenden, United Kingdom. It was at the Third International Symposium held in 1955 that the Society of European Nematologists (SEN, today named European Society of Nematologists or ESN) was founded. One year later, the first nematology journal, *Nematologica*, was published.

Although Western Europe had many plant-parasitic nematode species in common with the United States, the situation in the two regions differed. The cool European climate favored species such as the potato cyst nematode, the golden nematode and the sugar beet cyst nematode. In America, however, the major problem was the root-knot nematode.

In order to address nematode problems specific to the United States, nematologists participated in the Nematology Workshops sponsored by Shell Oil Company and the Technical Regional Committees during the 1950's. For years, nematologists primary venues for the regular exchange of information and paper presentations were at the annual meetings of the American Phytopathological Society (APS) and the American Institute of Biological Sciences (AIBS) symposia. At each meeting, between five to twelve papers were presented on the topic of nematology. W. H. Mountain, SON President (1964-1965), recalled, "in the late 1940's and early 1950's, we would find 10 or perhaps 25 people gathered in some quiet corner talking about nematodes, or what we thought we knew about nematodes. There would be the odd paper scattered through the meetings dealing with nematode problems. The plant pathologists might listen to these, but would ignore many of them because, you know in those days, most plant pathologists were quite certain that nematodes were not serious pests, except perhaps for the sugar beet nematode, the bulb and stem nematode, and maybe the root-knot nematode. And that would be a great big 'maybe'. But that group was the beginning of the Society [of Nematologists] we have today, and by 1952, at the [APS] Annual Meeting at Cornell University, we actually had a symposium on nematodes. A good many plant nematologists attended and listened. These informal gatherings became a regular thing at the APS meetings. They grew larger, and eventually entire sessions were set aside for papers on nematode [incited] diseases."

By 1958, a group of sixty-two individuals interested primarily in nematology attended the AIBS meeting at Bloomington, Indiana. These individuals met informally during the meeting and held late evening sessions to discuss nematodes and the future of nematology. During one of these informal sessions, a debate arose about the formation of a new society specifically focused on nematology. The discussion was divided between those who were strongly in favor of starting a nematological society and those vehemently opposed to the idea. D. J. Raski, in a letter to D. P. Taylor, wrote, "there were two alternatives which have been advanced to choose between, both of which have strong support. There are quite a few nematologists who believe we should resolve the needs of nematology within the framework of APS. There are also many who just as strongly believe we should proceed with steps to establish a separate society." Well-known nematologists such as J. R. Christie, B. G. Chitwood and A. C. Tarjan enthusiastically supported the venture of forming a new society, while others such as G. Thorne, J. P. Hollis and H. J. Jensen had reservations.

Those who objected to forming a new society for nematologists argued that most of the potential members already belonged to APS and gave their allegiance to APS. In addition, they felt that the younger potential members would not be able to assist adequately in the founding of a new professional organization and that the cost of starting a journal would be too great. G. Thorne commented, "we have neither the trained scientific workers nor the financial backing to justify a venture of this magnitude." The primary concern raised by the objectors was the sense of indebtedness to APS for its active nematology committee. W. B. Mountain remembered, "many of us felt that it was still too early. We had too few potential members. In addition, the subject matter committees of the American Phytopathological Society, including a vigorous Committee on Nematology, were reorganized and strengthened about that time."

In contrast, K. R. Barker observed that many members wanted to split from APS because "nematologists felt like second-class citizens." R. D. Riggs stated, "there was a need for a separate group from APS because they did not feel [the nematologists] were being represented sufficiently in what was already available." Individuals who supported the formation of a new society felt there was a need to split from APS, which focused primarily on plant-parasitic nematodes and the diseases they caused, to a new organization that would broaden the work and focus of nematology. J. N. Sasser recalled, "plant pathologists, even though they had considerable training in nematology, were in a plant pathology setting. There was resistance there. I do not know exactly why. I guess some of the plant pathology departments welcomed nematology as a discipline but did not want to see us go into another society. This was a natural feeling since it meant breaking away from the plant pathologists' society [APS]."

Although there was differences of opinion about forming a new society, those present agreed that it was a huge venture and should only be undertaken when there was a good chance of success. After considerable debate, the vast majority of the group participating in the discussion voted to authorize an Advisory Committee on Nematology, to consult informally with various APS council members and other colleagues to determine how best to meet the needs of the nematologists. The committee was comprised of: D. J. Raski, W. F. Mai, M.

**Figure 4-1.** Committee of Four. From left to right: D. P. Taylor; F. E. Caveness; O. H. Elmer and A. L. Jenkins.

B. Linford, W. J. Martin, H. J. Jensen and J. N. Sasser (chair). This group was to prepare a report to be presented formally at the 1959 APS annual meeting held at Pennsylvania State University.

Prior to the 1959 APS meeting, the Advisory Committee on Nematology published the results of their investigation in the Nematology Newsletter in order to give the participates time to carefully review the report before the meeting. The Committee reported that council members of APS were aware of the problems and issues raised by the nematologists and they wanted to find a solution. The report revealed the desire of APS council members to retain nematologists within APS rather than having them form a separate society. They did not envision any advantage, except autonomy, that nematologists would gain. The conclusion of the Advisory Committee was "that eventually nematology will be identified as a separate science and that the formation of a society of nematology is inevitable, a proposal for a separate society at a more appropriate time would have the blessings of all concerned."

One nematologist in particular, D. P. Taylor, believed that advancements in the science of nematology would only be insured if a new society was formed. W. B. Mountain recalled, "nothing daunted Don Taylor and a very few other individuals of similar dedication or stubbornness in their attempts to get this Society going." D. Taylor, nematologist, University of Minnesota, was joined by F. E. Caveness, nematologist, North Dakota State College; A. L. Jenkins, nematologist, University of Missouri and O. H. Elmer, nematologist, Kansas State College, forming the "Committee of Four". Its purpose was to establish an independent nematological society.

As the Advisory Committee initiated efforts to accommodate nematologists belonging to APS, D. P. Taylor contacted seventeen of his colleagues to inform them that a society of American nematologists was being established. In a letter dated August 22, 1959, D. P. Taylor wrote, "a society of American nematologists is to be established on or before January 1, 1960. Letters of invitation are to be sent to more than 300 potential charter members...At the forthcoming AIBS meeting at Pennsylvania State University [in December 1959], an announcement of the plans to organize the society will be made...If you plan to attend these meetings you are invited to speak in support of these efforts."

During 1959, nematologists were meeting at Pennsylvania State University in August with APS and in December with AIBS. In late August 1959, nematologists and plant pathologists, attending the annual APS meeting, filed into Room 111 of Boucke Hall at Pennsylvania State University. Greeted by J. N. Sasser, W. F. Mai, M. Linford, W. Martin, H. Jensen and D. Raski, individuals listened to the presentation of the Special Report by the Advisory Committee of Nematology. The report was accepted by an overwhelming majority attending the meeting. Letters submitted to the Committee by nematologists unable to attend the meeting also indicated overwhelming support for the creation of a nematological society.

After the conclusion of this meeting, however, it

became apparent that many in attendance had interpreted the report in "innumerable and variable ways". For the purpose of clarification, a new committee was established on September 1, 1959 outside the framework of APS. This committee of nine members was appointed to study the necessary steps for the formation of a new society. Their task included exploring the need for a separate society, examining the relationship of the new society with other scientific societies as well as developing a questionnaire to be sent to all those working in nematology to ascertain their opinion on the formation of a nematological society. This Committee of Nine was comprised of: D. J. Raski, University of California-Davis (chair); A. D. Baker, nematologist, Canada Department of Agriculture; J. R. Bloom, Pennsylvania State University; E. J. Cairns, Auburn University; D. A. Slack, University of Arkansas; A. C. Tarjan, University of Florida; A. L. Taylor, USDA at Beltsville, Maryland; D. P. Taylor, University of Minnesota, and J. D. Tiner, Rutgers University.

The appointment of this committee seemed to give legitimacy to the formation of a new society. A number of nematologists who originally opposed the idea began to support the endeavor. The time now seemed right for this handful of scientists to break ground and form a society of nematologists. As Sasser wrote to D.P. Taylor in the spring of 1959, "I am sure that once a majority as a group has decided that this is what we should do, then all will unite their efforts and do all possible to make it a successful society."

D. P. Taylor worked quickly to catapult the new Society into action. Just two weeks after the 1959 APS meeting, he engaged the Committee of Four and seventeen other nematologists to develop a list of potential members for the new society. On September 15, 1959, Taylor wrote to nematologists in Canada and the United States informing them that "a society of 'Nematologists of the Americas' is now being organized and is endorsed by the persons whose names appear below. Being an active worker in nematology you and your associates are invited to become charter members of this society. One of the principle objectives of this society is the publication of a journal devoted to all phases of nematology...With the consent of Dr. D. A. Slack, Editor of the Nematology Newsletter (NNL), NNL will become an official publication of the Society of Nematologists of the Americas...Dues will be $3.00 per year...Officers of the society will be elected from the total membership as of January 1, 1960 by mail ballot from a list of the five names most commonly suggested for each office by the membership. Subsequent elections will be made as defined in the constitution and bylaws." This letter included a membership application form.

Working separately, the Committee of Nine also compiled a list of nematologists working in the United States and Canada and created a questionnaire that was mailed to almost three hundred nematologists. The questionnaire was accompanied by a letter from the committee stating that "it is hoped that the attached questionnaire will reflect the wishes of the nematologists and serve as an effective guide for any plans that might result from the answers. If an organization is requested, this committee will proceed to draw up a constitution and organizational structure defined as completely as possible with a specific procedure for implementation."

The questionnaire was oriented to assess the specific areas of interest in nematology (e.g., plant-parasitic, soil, aquatic, invertebrate, vertebrate), the desirability of forming a nematological organization, the level of interest in joining the organization if it was developed, the type of publication desired as an outlet for reporting research data and the incorporation of the Nematology Newsletter as an instrument for the new organization. By the February 15, 1960 deadline, 171 responses were received by the committee. The results pub-

**Figure 4-2.** Committee of Nine. Starting from top left to right: D. J. Raski; D. P. Taylor; A. D. Baker; J. R. Bloom; E. J. Cairns; D. A. Slack; A. C. Tarjan; A. L. Taylor and J. D. Tiner.

lished in the March 1960 issue of the Nematology Newsletter were, "of more than 260 questionnaires distributed, a total of 171 were returned...The professional fields of interested listed by the number of pollees were as follows: 106 nematology, 58 plant pathology, 14 parasitology, 9 entomology, 1 microbiology and 1 horticulture. The work area in nematology designated by the number of pollees was: plant-parasitic nematodes-153, soil nematodes-55, aquatic nematodes-18, nematodes of invertebrates-20 and nematodes of vertebrates-17...In reply to the question 'do you desire the formation of a separate organization of nematologists?', 115 marked 'yes', 41 marked 'no' and 15 failed to indicate a choice. Of the 156 answering, however, 142 or 91% indicated they would associate themselves with the organization if it was formed. In reference to the type of organization desired, 125 individuals marked a choice: 107 preferred formation of a society, 7 preferred sectional status within the American Phytopathological Society, 8 preferred a club and 3 suggested other alternatives."

Concerning the publication of a journal, the vote was divided between starting a journal and utilizing existing nematological publications. However, 142 responders favored utilizing the Nematology Newsletter until a formal publication was established.

The Committee of Nine concluded that the survey clearly indicated a preference for the formation of a separate and new society. D. J. Raski commented to the Committee that "it seems to me quite clear that the wishes expressed through these questionnaires point toward the formation of a society and that we should proceed with this as soon as possible." A. C. Tarjan also addressed the group with the concern that "it is my opinion that our greatest danger is inactivity."

Inactivity would not be a long lived concern. Immediately, the Committee began to consider five important points in the formation of the organization: the society name, a constitution, publication of the organization, annual meetings and affiliation with other organizations.

The Committee believed that the first important step in the formation of a new society was to draft a constitution to establish the purpose, define objectives and formulate a structural foundation within which to operate. D. Taylor, who had already drafted a constitution prior to 1959, began to circulate it amongst the potential members for review and revisions. By the early part of the 1960's, the Committee of Nine had drafted the constitution and bylaws. Copies were circulated to all the members of the Committee of Nine as well as a number of major centers of nematology for input. The draft of the constitution was accompanied by an appeal for nematologists to join the new society.

In November 1960, two major points still remained unresolved. The first issue was the name of the society. The second was to affiliation with APS. To resolve the name issue, a proposal was made to decide the name by polling nematologists. The second issue of affiliation was resolved by including a provision in the constitution giving the ultimate responsibility of the decision to the governing body of the new organization. The constitution would continue to undergo revisions for the next two years.

While the constitution was in the process of development, D. Taylor was chosen as acting secretary for the new society until officers were elected. He and the Committee had been active in circulating copies of the constitution as well as mailing membership applications to nematologists. By February 1961, 150 membership pledges had been received.

As 1961 progressed, D. Taylor wrote to the new members requesting nominations for officers for the new society. Numerous individuals were nominated. The three nominees who received the largest number of nominating votes for the Presidential and Vice Presidential seats were D. Raski, G.

Thorne and A. L. Taylor. However, all three withdrew their names from the list. The remaining nominations were counted and the following people were placed on the First Executive Committee ballot:

>President:
>>M. W. Allen, University of California
>>J. R. Christie, University of Florida (retired)
>>W. F. Mai, Cornell University
>
>Vice-President:
>>E. J. Cairns, Auburn University
>>W. R. Jenkins, Rutgers University
>>J. N. Sasser, North Carolina State College
>
>Treasurer:
>>V. H. Dropkin, USDA
>>A. M. Golden, USDA
>>S. A. Sher, University of California
>
>Secretary:
>>D. A. Slack, University of Arkansas
>>A. C. Tarjan, Florida Citrus Experiment Station
>>D. P. Taylor, University of Minnesota

Of the approximate 150 issued, 135 ballots were returned and the results were:

President:
- M. W. Allen 57
- J. R. Christie 51
- W. F. Mai 27

Vice-President:
- J. N. Sasser 64
- E. J. Cairns 38
- W. R. Jenkins 33

Treasurer:
- V. H. Dropkin 56
- S. A. Sher 44
- A. M. Golden 34
- no vote 1

Secretary:
- D. P. Taylor 67
- D. A. Slack 39
- A. C. Tarjan 28
- no vote 1

**Figure 4-3.** First Executive Committee for the Society of Nematologists in 1961. From left to right: Merlin Allen, President; Joe Sasser, Vice President; Victor Dropkin, Treasurer and Don Taylor, Secretary.

On June 26, 1961, D. P. Taylor reported the result of the election of officers for the new society. Individuals declared elected for the respective positions were: M. W. Allen, President; J. N. Sasser, Vice President; V. H. Dropkin, Treasurer; and D. P. Taylor, Secretary. J. N. Sasser recalled, "I will never forget when the Society was formed. They elected officers. Merlin Allen was elected as the first President and I was the first Vice President. I remember going to my department head's office, Don Ellis, and saying, 'look, Dr. Ellis, they have elected me the Vice President of the Society of Nematologists. I feel like I ought to support it, go to their meetings and contribute to it.' He said, 'absolutely. That is your duty. That is your job.'"

The new President announced upon his election that the first item of business would be to ratify the proposed constitution and select and approve a name for the new society. According to the constitution drafted in 1959, the new society would tentatively be called "The Society of North American Nematologists" until an official society name was designated. During the election process of new officers, Taylor had asked all the pledgees to suggest a name for the new society. Three names received a majority of support: American Nematological Society, American Society of Nematologists and Society of Nematologists. The new officers considered the names and decided that the new society should be open to all nematologists around the world. The new organization was therefore titled the Society of Nematologists (SON).

With an official name chosen, the next item of business was the ratification of the constitution and bylaws. On February 1, 1962, copies of the constitution along with a letter from M. W. Allen urging the members approval were mailed to 155 members who had paid their dues and 27 who had pledged to be members. By March 14, 1962, 121 ratification blanks, 66.12% of the forms, were returned. According to Article VIII of the Constitution, "ratification requires approval by two-thirds of the total membership." Under this rule, ratification was one vote short. However, Article III defined a member as "a person who pays dues as prescribed in the bylaws." Based on this, there were only 155 members and of those, 107 ratified the constitution. This made up 69.03% of the official membership. Thus, it was concluded that the Constitution had been officially ratified pending the officers approval.

With the election of officers in January 1961, the first meeting of the Executive Committee in late December 1961, the ratification of the constitution in early 1962 and a total of $482.77 in cash assets, the Society of Nematologists had become a reality.

*I believe the underlying philosophy of our new society is that the <u>nematode</u> is the hub of the wheel, that the spokes are the various subject matter areas which must be studied and advanced if we are to understand the "hub".*

*-President J. N. Sasser*
*SON Presidential Address*

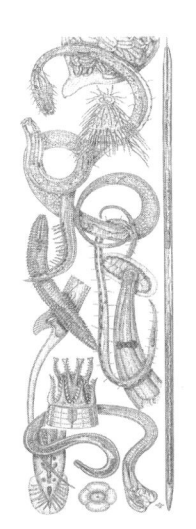

# 5 The Golden Age of Nematology

At the outset of the 1960's, the Society of Nematologists was born and for the next serval years it underwent continuous growth and change. The goal, to join together all aspects of nematology into one comprehensive discipline, was beginning to be realized. As W. B. Mountain stated, "the object of our Society is the same as that of any other learned society. It is to further the advancement of a science, in this case nematology. Its main function isn't to hold annual meetings or to bring the members together in one way or another. Its function is to serve the entire discipline. It is to act as the skeleton around which the discipline grows and develops. It is to provide strength and meaning and form to an area of learning."

With the birth of SON, nematology was receiving more recognition as a separate and vital entity. Nematologists worldwide began to come together to work for a common goal, the advancement of the science of nematology. G. Thorne remembered, "perhaps no other science has enjoyed the enthusiastic reception as that accorded to nematology during the past 20 years. Where once it was regarded with skepticism and contempt, it is now accepted as a necessary part of training for work in the field of agriculture...From the few part time workers in 1917, the work force here in the United States has increased to about 125 full time employees who are delving into the various phases of nematology."

In the 1960's, the science of nematology grew at a rapid pace. Departments or divisions of nematology were being formed and plant pathology departments and some entomology departments were creating positions for full time nematologists. Nematological awareness increased to such a point that an article about nematodes was published in the Reader's Digest. J. N. Sasser recalled, "it was a tremendous article on the importance of these nematodes. It gave everyone a boost. The dedication, enthusiasm and expertise of all the young scientists getting into the field at that time was just phenomenal."

Besides the enthusiasm generated by these energetic nematologists, other factors also contributed to this "Golden Age" of Nematology. The 1960's, like the 1950's, was a time of accelerated population growth. However, the population demographics were shifting. In the 1960's, 9% of the US population lived on farms and only 8.3% of the population were farmers compared to the 1950's when 16.5% of the population lived on farms and 12.2% were farmers. This fact combined with industrial growth and the expansion of the ground transportation system made the need for sustainable agriculture more apparent. The government responded to these problems by passing legislation for the preservation of farm land and in 1968, the government provided four million dollars for continued research in the discipline of nematology.

These factors played a large role in the growth of the science of nematology as well as in the Society of Nematologists. Not only was funding pouring in from the government but commercial agricultural companies began to sense the importance of nematology. Company representatives regularly attended the Society's annual meetings and provided financial support to sustain the organization. The companies that became sustaining members during the 1960's included: Chevron Chemical Company, Dow Chemical

Company, Eli Lilly and Company, Mobil Chemical Company, Morton Chemical Company, Niagara Chemical, Shell Chemical Company, Union Carbide Corporation, Velsicol Chemical Corporation and Vero Beach Laboratories, Inc. This tremendous support was logical since chemical control of nematodes was an important method at that time. J. N. Sasser stated, "at the time of my presidency, our main weapon for control was chemicals...These companies were trying to sell their products...we were in it together."

Due to the rapid growth and notoriety of SON, the new Society became the center of conversation at the APS meeting held in Biloxi, Mississippi in December 1961. The Nematology Committee of APS organized a very strong program at the meeting including an impressive symposium on interactions of nematodes with other organisms causing plant disease. It was at this meeting that the newly elected SON Executive Committee conducted their first Executive Meeting on December 11, 1961. Still working out the finer details of their new society, V. H. Dropkin, C. W. McBeth, J. N. Sasser and D. P. Taylor took this opportunity to discuss such issues as the constitution, the term of office for the Executive Committee, the publication of a journal, membership, membership dues and the SON letterhead for the Executive Committee. The two main issues the officers addressed were affiliation with APS and AIBS and the SON annual meeting. The issue of annual meetings was resolved by making plans to conduct the first Society of Nematologists Annual Meeting at Corvallis, Oregon in 1962 but the topic of affiliation remained unresolved. It was also during this APS meeting that the new members of SON conducted the First Business Meeting.

The First Annual Meeting of the Society of Nematologists was held at Oregon State University, Corvallis from August 26-29, 1962. The Society met as an affiliate society of AIBS and conducted joint sessions with the nematology section of APS. The first SON annual meeting program consisted of the following events:

> Sunday, August 26-Room 153 Physics Chemistry Building
> 7:30 p.m. Joint Meeting with Nematology Section, APS
> Monday, August 27-Sackett Hall, B
> 1:30-4:30 Joint Meeting with Nematology Section, APS
> Tuesday, August 28-Sackett Hall, B
> 10:15-12:00 Joint Meeting with Nematology Section, APS
> 2:30-5:00 Society Executive Committee Meeting-Room 300 Withycombe Hall, M. W. Allen, presiding
> Wednesday, August 29-Room 209 Withycombe Hall
> 1:30-3:30, M. W. Allen, presiding
> Guest speaker-Dr. Ellsworth C. Dougherty, Department of Nutritional Sciences, University of California, Berkeley. "The New Nematology-Its Coming Impact in Biology?"
> A business meeting of the Society followed this address.

The Executive Committee decided that no formal paper sessions would be held by the Society of Nematologists at this meeting. However, papers on nematology were scheduled by APS in the nematology sessions. One could argue that the tradition of the SON Banquet began in Corvallis when Dr. and Mrs. H. Jensen invited members of the Society to a social evening and buffet supper at their home.

Although the First Annual Meeting of the Society of Nematologists was a success, there still remained one major concern, affiliation with other societies. The question of a joint or an independent meeting continued to be a topic of debate. Following the Second Annual Meeting held at Amherst, Massachusetts in 1963, discussions on affiliation with other societies were intensive. M. W. Allen, first SON President (1961-1963), expressed his concern over the issue that SON would not meet jointly with APS. He stated, "the majority of [SON] members expressed their desire to meet at close intervals with APS in a poll a few years ago." However, W. B. Mountain, SON President (1964-1965), and others argued that interest in frequent joint meetings with APS had changed after the formation of the Society. One of the reasons for this change was the rapidly increasing number of new members. Mountain asked, "do we really want a Society of Nematologists-that is a Society that has real meaning and serves a real function-or do we want the old group meeting together at the APS meetings? Or, do we really want some type of combination of these two? But is there any choice? Don't we require either a separate, autonomous society, or no society at all?"

In an effort to resolve this issue, the Executive Committee was asked to give their opinions on the topic. W. R. Jenkins, nematologist, Rutgers University, wrote to J. N. Sasser expressing that "there has been a lot of discussion among various persons concerning the Society of Nematologists. As you know I am a member of the Nematology Committee of the American Phytopathological Society. It seems that the opinion of this Committee in general is that the Society of Nematologists should begin a vigorous program separating itself more completely from APS. These are certainly my personal wishes as I have a great interest in seeing nematology become identified in itself." D. A. Slack also supported an independent meeting. He stated that "SON must develop and pursue its own program separate from APS if it is to function and serve nematologists as a society independent of APS. A decision must be made as to when SON will assume this responsibility." I. J. Thomason, SON President (1975-1976), commented, "I personally feel that SON is leaning too heavily on APS. As I recall early discussions leading to the formation of SON, 'autonomy' was a frequently used word."

**Figure 5-1.** First Annual Meeting of Society of Nematologists at Oregon State University, Corvallis, Oregon 1962.

After listening to the advice of the Executive Committee, Sasser informed the Nematology Committee of APS as well as the SON Executive Committee that "the feelings were almost unanimous that the Society of Nematologists should meet separate from APS in 1964 and meet with AIBS at Boulder, Colorado...Merlin Allen also authorized me, acting for the Committee, to officially announce that our 1964 meeting would be with AIBS. Accordingly, I will prepare a notice to appear in the next issue of NNL." The Third Annual Society of Nematologists Meeting would be the first independent meeting held by SON.

Hosting annual meetings independent of APS entailed additional responsibilities for SON. One major issue was the publication of abstracts presented at the annual meeting which could no longer be published in *Phytopathology*. Thus, an outlet needed to be found. It was suggested that *Nematologica* be the medium for publication until a scientific journal was established for the Society of Nematologists. This need for printing meeting abstracts fueled the urgency for a SON publication. A journal for SON would begin to be seriously evaluated during 1965 and 1966.

The Society had already begun publishing news in the Nematology Newsletter (NNL), an instrument that became an official organ of the Society of Nematologists in 1962. As a nematology news medium, the NNL began to be circulated worldwide. V. H. Dropkin commented, "we are proud to have this Newsletter as the official 'voice' of our Society." J. N. Sasser began to use the NNL as a vehicle for the annual meeting announcements and information. He stated that the "proper function of the NNL is not only to publicize annual meetings in advance but also to report to Society members the details of outstanding events which occur at the meetings. This gives all members ready access to this information and will surely serve to strengthen the Society by giving members tangible return for their financial support." The first official Society of Nematologists Membership Directory was published as part of the Nematology Newsletter [Volume 8(3)] in 1962.

Although the Society had this publication at its disposal, the need for a journal was still pressing. J. M. Ferris, NNL co-editor (1963-1965), stated that the "NNL is not a scientific journal, and we do not intend to try to make it one." V. R. Ferris, NNL co-editor (1963-1965), upheld this position by commenting, "we have never at any time conceived of the Nematology Newsletter as being or becoming a standard journal for publication of original research."

From 1965 to 1966, G. C. Smart researched the feasibility of publishing a journal for the Society of Nematologists. The Executive Committee had previously proposed the idea of an annual journal for the purpose of publishing papers presented at the annual meetings. Smart personally favored this idea. He proposed, however, that the publication not only be used for the publication of meeting abstracts but also as an outlet of nematological research to be printed several times a year. The idea was considered and Smart was asked to develop and distribute a questionnaire to SON members to determine interest in the proposed publication.

The questionnaire contained questions such as: "Do you, in principle, favor publication of a journal?"; "What year would you prefer to see publication begin?"; "Based on our present size and financial status, how often would you prefer publication?"; "Would you be willing to pay if necessary?"; "Would your institution pay a page charge for you to publish?"; and "Would you be willing to serve, if asked, as Editor-in-Chief or on the Editorial Board?"

Publication of a journal by SON received strong support based on those who responded to the survey. In Smart's report to the Executive Committee, he indicated that "those responding overwhelmingly favored publication, in principle, at least, and the majority wanted a journal as soon as possible. While the majority favored a quarterly publication, it may well be that at this time a less frequent issue will have to be considered due to lack of finances...I was very happy to note that the majority of those answering are willing to work on a journal. While no one was brash enough to state that he would serve as Editor-in-Chief, only 14 stated that they would serve either in that role or on the Editorial Board, and 36 would serve on the Editorial Board." Smart urged that "whatever we decide to do though, I feel that we must be ready to begin our journal by the August 1968 meeting because it seems very likely that we may not have a place to publish abstracts of our papers at that time- if we desire them to be published and we certainly will not have anywhere to publish symposium papers if we hold one. Therefore, the time to act is right now!"

Following Smarts suggestion to "act now", the Executive Committee met at Washington D.C. in 1967 to discuss results of the survey and the possibility of developing a scientific journal. Following a lengthy deliberation, the Committee decided to initiate the publication with a biannual publication. One issue would be dedicated to the papers presented at the annual meeting and the second issue would be dedicated to the publication of other manuscripts. V. G. Perry, SON President (1966-1967), commented that "we finally settled on a format for JON which was patterned after the Florida Entomological Society Journal." The decision that the Society would publish a scientific journal was announced at the Sixth Annual SON Meeting during the Business Meeting at Washington, D.C. G . C. Smart commented, "the decision by the Society of Nematologists to publish a journal is probably second only to the decision to begin the Society of Nematologists." The NNL reported that the Journal published by SON would "become the first and only one devoted exclusively to nematology and published by a scientific society."

With the decision to publish a journal, the process for developing the journal began but it would be 14 months before the first issue was published. The Editorial Board of the new journal met on November 1-2 ,1967 at Beltsville, Maryland to discuss the format, subject matter, publisher, finances, editor-in-chief and a number of other issues related to this new project. S. D. Van Gundy, nematologist, University of California-Riverside, was unanimously elected as the first Editor-in-Chief. At this meeting, *Journal of Nematology* was selected as the official title for the publication. Although the Executive Committee at the Sixth Annual Meeting had approved a biannual publication, the Editorial Board felt that it was necessary to publish a quarterly journal. The Board decided that only papers dealing with original research from basic, applied, descriptive and experimental nematology would be considered for publication. Van Gundy stated, "[the *Journal of*] *Nematology* will begin publication as a quarterly journal with Volume 1, No. 1 scheduled to appear in January 1969. It will serve as a medium for the publication of original research in all fields of nematology." The cost of the Journal would be covered by an increase in annual membership dues from $3.00 to $6.00. Along with this announcement, the Editorial Board sent out a general invitation to all SON members to submit manuscripts to be considered for publication.

Although the stage was set for the Journal to be published, not all aspects had been resolved. The cover design as well as a logo design for the Journal were major points of debate. In order to decide upon a cover, informal polls were taken at several nematology laboratories. Four cover designs and four different colors were presented to selected members of the Society for their opinion. One design used the Journal's table of contents on the front cover (Design A). The second cover was a design by N. A. Cobb and W. E. Chambers for a journal they planned to publish but never accomplished (Design B). A third cover design was a nematode overlay taken from original sketches by Charles Papp and S. A. Sher (Design C). The fourth cover was modeled after the journal *Nematologica* (Design D).

Determining the cover design and cover color was a topic of lively discussion. A straw poll taken by V. R. Ferris, SON Secretary (1965-1968), in her laboratory at Purdue University demonstrated the variation in opinion. She stated, "half of the 6 people queried felt that the "B" design should be used for the first year for sentimental reasons. One thought it should be used permanently! Three people voted for design "C"; one for cream [cover color] and two for green. Two people voted for design "A"; one for cream and one for green...I personally voted for the "A" design (in cream) because I find it a convenience to have the contents right on the cover." S.

Van Gundy stated, "everyone thought the orange color was good to single out the journal on the book shelf." After considering all of the feedback, the Editorial Board decided that the official cover for the new *Journal of Nematology* would be the sketches by Charles Papp and S. A Sher (Design C). The orange colored cover was ultimately chosen by the Board. This color would become one of the major identifying marks of the Society.

The Editorial Board decided to dedicate the first issue of the new journal to N. A. Cobb. The decision was in "recognition of his pioneering work in nematology. He probably has had a more profound and permanent influence on the profession of Nematology than any other individual...The Society of Nematologists and the Editorial Board hope this journal will serve to develop and to promote the Science of Nematology so ably initiated by Dr. Cobb in the early 1900's." The first issue of the *Journal of Nematology* was published in January 1969. It was praised by nematologists as being a high quality journal. A. M. Golden telegraphed S. Van Gundy stating, "congratulations on a job well done. JON is beautiful."

Another landmark event took place in 1969 when members of the Society of Nematologists were asked to vote on an official logo for the Society. Two designs were published in the Nematology Newsletter along with a request for members to indicate their preference. The membership chose the design which showcased the Western Hemisphere in the center region of the globe [Design A]. The Executive Committee accepted the results of the vote and that symbol became the official logo of SON in 1969.

The ten years spanning from 1960 to 1970 marked huge strides for nematologists. The science received enormous support from the government, agricultural companies and fellow scientists. Perhaps the largest support came in the form of the newly founded Society of Nematologists. Within

48   Agriculture, Nematology and SON

**Figure 5-2.** Proposed covers for the *Journal of Nematology*. Each cover was proposed in four different colors: orange, green, blue and cream.

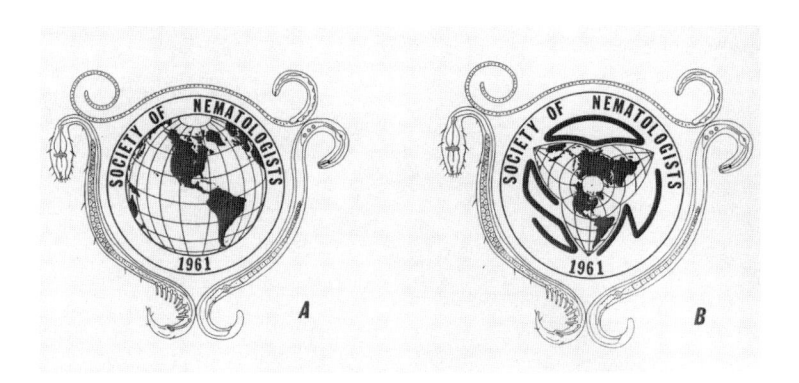

**Figure 5-3.** Two proposed SON logos that were voted on by members in 1969.

less than ten years, the Society of Nematologists was organized, membership formed, an executive committee elected, its constitution and bylaws were in place and the *Journal of Nematology* was published. These "golden" years formed the strong root system that would support and nourish nematologists during the upcoming years of change.

*Reflecting on the changes in nematology during the past 50 years, from using mechanical calculators and typewriters to high-capacity computers for data management, development of nematode-host-environment stimulation, basic physiological work to molecular biology and related analytical equipment, including development of DNA probes, cloning genes, engineering host resistance, and genomics, ecology-based nematode-IPM-crop management to using the Internet to instantly sharing information worldwide - is sometimes mind boggling.*

-President K. R. Barker
SON President

# 6 The Evolution of Nematology

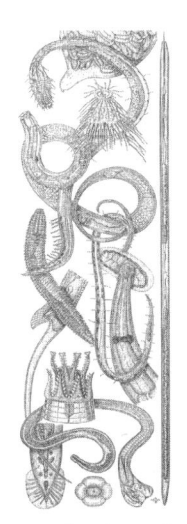

The dawn of the 1970's saw a continuation of the agricultural prosperity experienced in the previous decades. Nearly ten million Americans lived on farms in the early 1970's comprising 4.9% of the labor force. Although the percentage of farmers was lower than that of the 1960's, the US population experienced a slight migration reversal. During this era, the trend toward urbanization slowed as more individuals moved to rural areas. This increase in farm activity was due, in part, to the money made available through farm loans which were intended to expand farming operations. In addition to the slight increase in the percentage of people living in rural areas in the 1970's, there was a modest increase in the average farm size which grew from 390 acres to 426 acres. With the increase in operations, crop yields began to exceed the needs of the US market. The surplus generated from the farm expansion was absorbed by export, especially of grains and oilseeds.

The agricultural prosperity of this era resulted in major advances in agricultural technology. Agricultural research received global notoriety in 1970 when Norman Borlaug was awarded the Nobel Peace Prize for developing semi-dwarf, high-yielding, disease-resistant wheat varieties. This scientific breakthrough not only heightened research awareness, it also had a major impact on farm production. For example, by

1963, 95% of Mexico's wheat fields were planted with semi-dwarf varieties developed by Borlaug. In 1963, the harvest was six time greater than in 1944.

Agricultural technology was evolving and the major emphasis was shifting to molecular biology. Having been a vibrant field of research in the 1950's and 1960's, technological advances and new developments in the 1970's provided molecular biology with the tools needed to make major advances. D. Bird, molecular nematologist, North Carolina State University, commented that "the invention of molecular cloning built around discovering restriction enzymes, the use of electrophoresis to analyze DNA and the development of DNA sequencing by Paul Berg in 1972 have provided us with tools to really empower our discipline."

Molecular biology began to impact nematological research in 1974 with Sydney Brenner's molecular and developmental biological research done on the nematode species *Caenorhabditis elegans*. With these advances, H. Ferris, nematologist, University of California-Davis, stated, "we moved into this era of new technology in biology. Advances in any science are determined by the advances of the technologies that applies to them...some new technology helps knowledge accelerate. We had the fact that Brenner selected *C. elegans* as the model for understanding the developmental biology for the nematode genome and then the human genome cascaded from this."

Despite the technological strides in the field of nematology, nematodes continued to be controlled predominantly by fumigation. S. Lewis, nematologist, Clemson University, remembered that in the mid-1970's, "we still had dibromochloropropane [DBCP] that could be applied any time-before planting and after planting. It was an amazing chemical that was very efficacious. That was the standard. It was such an important part of the agriculture at that time. We did not really get into rotations and biocontrol because the chemical was so good. That is what most [growers] were using. The fact that you could use it on, say, live peach trees, was excellent. Chemicals were very important." However, the heavy reliance on nematicides was about to change dramatically, having a permanent affect on the science of nematology.

The later part of the 1960's and early 1970's was an era of environmental awakening. Many individuals began to openly discuss the overuse of pesticides and the damage they were causing to the environment. University students closely followed these debates and were greatly influenced by individuals such as ecologist Rachel Carson. Carson, in her widely read work, *Silent Spring*, warned the public of the effects of long term use of pesticides in agricultural practices as well as in scientific research.

Her work, along with the research of many others, increased the awareness of groundwater and surface water pollution throughout the world. Consequently, pesticide residues in food and the potential health risks associated with consuming contaminated food and water became major issues. In response to public concern about contaminated ground water, the Environmental Protection Agency (EPA) enacted the Federal Water Pollution Control Act Amendment of 1972. The Act was amended in 1977 and became commonly known as the Clean Water Act. This Act empowered the EPA to establish the basic structure for regulating discharges of pollutants into the waters of the United States. It gave the EPA authority to implement pollution control programs such as setting wastewater standards for industry. The Clean Water Act also continued to set water quality standards for all contaminants in surface waters. The Act made it unlawful to discharge any pollutant from a point source into navigable waters, unless a permit was obtained under its provisions. It also funded the construction of sewage treatment plants under the construction grants program and addressed the critical problems posed by non-point source pollution.

The Clean Water Act directly impacted nematology and many other agricultural sciences. Ground water contamination by pesticides including nematicides resulted in the prohibition or restriction of numerous chemicals from being sold on the open market. The EPA suspended the use of 1,2-dibromo-3-chloropropane (DBCP) in most of the USA in 1977 and permanently banned the chemical in 1979. An exception was granted for pineapple growers in Hawaii (the only exception to the ban in the USA) but this exception was reversed in 1985 resulting in a total ban on DBCP. Ethylene dibromide (EDB) was also judged to be unsafe by the EPA and use was suspended. Soon thereafter, Shell Chemical Company removed its nematicide 1,3-dichloropropene,1,2-dichloropropane (D-D) from the market and it was no longer manufactured. With the declining market for nematicides, chemical companies were not realizing sufficient profits to continue their research and production of nematicides. The result was a significant reduction in the number of compounds available to growers.

The added concern of the societal impact of chemical control on the environment changed the approach to nematode management. V. R. Ferris stated, "although it is difficult to place priorities on the social needs for which research in nematology can be of benefit, certainly the goal of an increased food supply for a growing world population through control of parasitic nematodes must be near the top...The development of environmentally safe controls for nematodes is important. Toward this end, we must increase research on basic concepts which govern nematode interactions with their immediate environment."

Contrary to the 1950's and 1960's view that chemical pesticides and nematicides were a miracle of modern science, the dawn of the "green era" placed emphasis on the negative

Evolution of Nematology 55

**Figure 6-1.** SON's American Society for Testing and Materials (ASTM) Subcommittee Meeting at Riverside, California in 1974.

impact of these pest controls. The new era called for different methods of control that were effective and imposed minimal harmful effects to the environment and nontarget organisms. The term given to this practice was Integrated Pest Management (IPM). The basic concepts of IPM was to use a combination of pest management methods to reduce pest populations to or near non-damaging levels with minimal impact on the environment. One of the goals was to reduce the use of chemicals wherever possible. As a consequence, crop rotation and resistant varieties became important tools in the management of nematodes and other pests.

Different scientific disciplines, including nematology, were working together in an effort to promote Integrated Pest Management. The objective was to foster interdisciplinary activities in research, teaching and extension to promote the development and use of IPM. The scientists believed that this program would have favorable economical, ecological and sociological consequences.

By the late 1970's and early 1980's, farmers began

searching for more cost-effective ways of farming by substituting on-farm resources for chemical products. In response to farmers needs, created by the environmental issues of this era, scientists and government officials developed LISA (low-input sustainable agriculture). LISA was based upon the two concepts, low input and sustainability. Low input dealt with the reliance on resource conservation of a farm and less emphasis on utilizing chemicals such as commercial fertilizers and pesticides. Sustainability, on the other hand, focused on what M. G. Cook referred to as "farming systems that are capable of maintaining their productivity while providing benefits to society indefinitely. The term sustainability has agronomic, environmental, social, economic and political dimensions." Funding for sustainable agricultural programs received $4 million in 1987 and by 1991, $40 million was appropriated.

Even though the economy appeared strong in the 1970's and funding for university research was relatively abundant, there were times during the decade that plant pathology and other agricultural departments began to experience economic restrictions. One of the striking aspects was the small number of positions available for nematologists and plant pathologists. H. Ferris recalled, "it was difficult to get jobs in plant pathology and nematology in the 1970's. One of the concerns was that there were too many graduate students being trained and there would not be any jobs for them."

These economic restrictions continued and worsened in the 1980's. During this decade, the agricultural community witnessed farm bankruptcies, a surplus of agricultural commodities and decline in the US international trade balance. In 1981, the United States found itself in the midst of an economic recession and the agricultural community was profoundly affected. Farmers who were once making a good living could no longer sustain themselves by farming. During the 1970's, many farmers, through the aid of government loans, had expanded their farms on the belief that crop prices would continue to rise. However, with the decrease in crop prices and the high interest rates on farm loans, many farmers were forced into bankruptcy. E. Bouska, an Iowa farmer, recalled, "land prices were being inflated. Banks and lending associations were too lenient about giving loans. People had over extended themselves. None of them had the experience with hard times. They over extended and could not pull out."

Due to these hard economic times, many individuals left the rural areas. Farmers who were unable to pay off their loans sold their land. Some of the farm land was sold to individuals from Europe and Japan who were able to purchase significant acreage during this depressed economic time. Farm prices would continue to decline until 1987. The decrease in demand for American grown produce was furthered by an increase in competition with European and Canadian grain and animal exports. This was due in part to the US-Canada Free Trade Agreement (CFTA) in 1989.

Despite the hard economic times, the Society of Nematologists continued to experience phenomenal growth during the 1970's and 1980's. In 1970, the Society debated the need to incorporate SON. The central issue was the fact that under the Society's constitution, each member of the Society was liable if legal action was taken against the organization. W. F. Mai, SON President (1968-1969), commented, "I look upon incorporation of SON only as a means of meeting a legal technicality to protect the officers and other members of the Society." The resolution that was drawn up for incorporation stated, "whereas the Society of Nematologists is an ever growing and expanding society, and whereas the Society seeks to continue to expand and promote the knowledge and the advancement of the science of nematology, and whereas the Society is an international organization increasingly involved with programs concerning the advancement of the science of nematology, and whereas the members of the Society seek the benefits of corporate existence and recognition, now, therefore, it is resolved that the Society of Nematologists be incorporated in the State of Maryland as a non-stock corporation by the law offices of Green, Swingle, Dukes and Mann and resolved that the corporate name shall be the Society of Nematologists, Inc." The Executive Committee agreed upon incorporation and each charter member was asked to approve or disapprove the articles of incorporation and bylaws. On August 25, 1971 at the Annual Meeting in Ottawa, Canada, the articles of incorporation were approved and the Society of Nematologists was officially incorporated on June 23, 1972.

The Society of Nematologists experienced the height of its success from 1980-1985. During this time, the membership of SON peaked at 915 members. Involvement in SON during this period was also at an all time high. A. Nyczepir, nematologist, USDA, remembered, "when I was a graduate student and then Research Associate with the USDA, both Steve Lewis and John O'Bannon, respectively, were always big promoters of the Society. Both of these past presidents encouraged me to get involved with SON in some capacity. I just felt that I wanted to help out, so I initially got involved with different committees. One committee of particular interest was the Membership Committee. I also loved and looked forward to going to SON meetings, because even though it was a relatively small society, it was a close-knit group of people which made it so nice."

With the rapid growth of the Society many operational issues which had been overlooked required urgent attention. For twenty years, the Society's charter members had operated on verbal communications and memory. D. W. (Freckman) Wall recalled, "whenever we elected a president, there was nothing to fall back on in terms of how to do business. There was issue after issue. There was no documentation to fall back

Evolution of Nematology 59

**Figure 6-2.** SON Executive Board at Seventy-fifth Anniversary meeting of APS at Ames, Iowa, 1983. Top Left to Right: D. Dickson, J. Veech, G. Noel and G. Fassuliotis. Bottom Left to Right: J. Webster, D. Wall, A. Johnson and D. Hussey.

on." In an effort to solve this problem, the Executive Committee in 1980 appointed an ad hoc committee to establish an operation manual and revise the constitution and bylaws. D. W. (Freckman) Wall was appointed chair and fellow SON members, G. W. Bird, C. Heald and D. W. Dickson were appointed to the committee. The project was completed in 1982. The intent of this new manual was to have a document that could be used as a guildline for officers and committees. It was designed to be updated as changes occurred in business operations of the Society. L. Krusburg, SON President (1991-1992) stated, "our operational manual is a dynamic work. Items become obsolete; sections need to be changed; overlooked items need to be inserted; and new items need to be added."

Beside the development of an operation manual, an ad hoc committee chaired by K. B. Barker was also established to

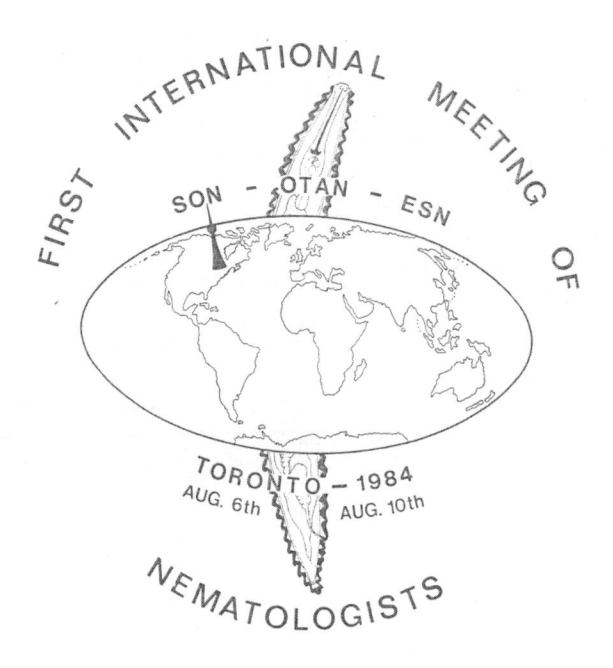

**Figure 6-3.** Logo designed for the First International Nematology Congress in Guelph, Canada in 1984.

update the constitution and bylaws of the Society. Revisions included changing the term "Executive Committee" to the "Executive Board" as well as adding a presidential office. For example, the presidential line of succession passed from Vice President, President to Past President. The new succession changed to include a new office. The success would now pass from Vice President, President-Elect, President to Past President.

At the same time SON was looking inward to strengthen its core, it was also reaching out to embrace the growing globalization of nematology. During the early part of the 1980's, the formation of an international organization of nematological societies to serve and foster nematology worldwide was conceived. A. C. Tarjan remembered, "the concept of

Evolution of Nematology  61

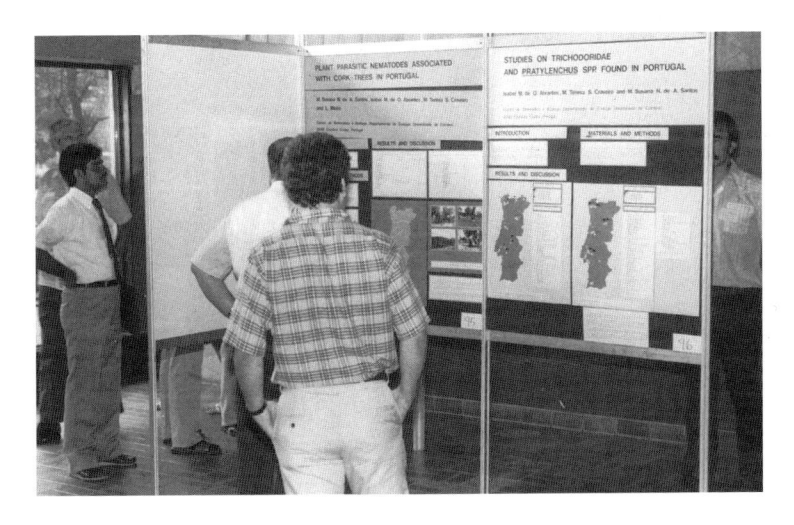

**Figure 6-4.** Poster Session at First International Nematology Congress at Guelph, Canada in 1984.

having an international organization in nematology was born in Vineland, Canada with Theo Olthof and his group there being the parents. Theo subsequently wrote to several people, among which was Professor Grossman in Germany who then was President of the International Society for Plant Pathology. John Webster, SON President (1982-1983), received copies of all the correspondence and was in close touch, and agreement, with Theo on the concept." K. R. Barker stated, "John Webster really had the vision and the foresight for a federation of nematological societies and the need for this sort of interface among the various small societies of nematology."

Working toward this goal, an ad hoc committee was formed with J. Webster, Canada; A. C. Tarjan, the United States; N. Hague, the United Kingdom; R. Brown, Australia; M. Luc, France; B. Weischer, Germany and A. Maqbool, Pakistan. The committee worked on a constitution, statutes and rules of procedures for the new organization. At this time, there were 17 nematological societies worldwide. In 1984, the

62   Agriculture, Nematology and SON

**Figure 6-5.** SON Silver Jubilee Celebration, Orlando, Florida, 1986. Past SON Presidents. Left to Right: A. M. Golden; V. R. Ferris; J. O'Bannon; B. Endo; D. Wall and R. Rodriguez-Kabana.

**Figure 6-6.** SON Honors and Awards Banquet at the Silver Jubilee Celebration, Orlando, Florida, 1986.

Society of Nematologists, Organization of Tropical American Nematologists (ONTA) and the European Society of Nematologists (ESN) sponsored the First International Nematology Congress (FINC) in Guelph, Canada. Deemed a "great success," participants agreed to continue developing an organization that would facilitate the holding of future meetings similar to the FINC. The different organizations believed that "the time has come to try to promote this specialty and to try to establish close contacts with societies of nematology." Consequently, future congresses were planned.

Although J. Webster wanted to promote the international congresses, his efforts were channeled into forming a formal organization to foster international cooperation amongst nematologists. J. Webster stated, "nematologists appear to want a federation of nematological societies rather than a new society." K. R. Barker observed, "the Americans and Canadians were very hopeful that from the very First Congress in Guelph that things [for an organization] were all set up...But everything was sort of given up until the meeting in The Netherlands at the Second International Congress."

As Webster and others were laying the groundwork for future collaborations, the Society of Nematologists celebrated its Twenty-fifth Anniversary in 1986. In commemoration of this milestone, SON decided to publish a collaborative work from nematologists as a recognition of notable scientists as well as a tribute to the Society of Nematologists. This publication, edited by J. A. Veech and D. W. Dickson, was entitled *Vistas on Nematology*.

At this time, D. W. Dickson, University of Florida, fostered the concept of a journal based on summary reports. SON made the major decision to publish such a journal, the *Annals of Applied Nematology*, devoted to applied research such as results of nematicide tests and variety screening trails. This journal would serve as an outlet for research data that would not otherwise be published. The *Annals* published research specifically related to management of plant-parasitic nematodes and survey data. J. L. Starr, JON Editor (2000-2003), stated that "as a peer-reviewed science journal, publication in the *Journal of Nematology* was based primarily on hypothesis testing and experimentation. The *Journal of Nematology* was an appropriate outlet for manuscripts on nematode management. It was not an appropriate outlet for manuscripts based on routine screenings of germplasm for host status or surveys of distribution of plant-parasitic nematodes associated with a particular nematode associated with a particular crop in a limited production region...whereas the *Annals of Applied Nematology* was an appropriate outlet for such manuscripts." The first two issues were published in 1987 and 1988 as volumes one and two; however, due to problems, subsequent issues were corrected and officially referenced as the *Supplement to the Journal of Nematology*.

As the Society was celebrating nematological advances a foreboding wind began to blow through the science. A major shift was about to take place in nematology that would have a dramatic impact on the science, scientists and the society of nematologists. Over thirty years had past from the time nematology emerged as a separate discipline and as the first generation left the field a new generation was waiting in the wings.

*Since many of today's nematologists were trained in the early 1950's, our Twenty-fifth Anniversary marks the approaching retirement of the bulk of our charter members...Thus, this anniversary marks a transition between the older era of nematologists and the new generation of members of our discipline. These new bright, young nematologists are being trained in other disciplines...They will provide the new direction in nematology.*
-S. D. Van Gundy
Vistas on Nematology

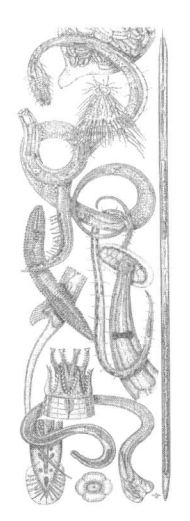

# 7 Dawning of a New Generation

For years, the science of nematology had experienced rapid growth and notoriety thanks to the tireless efforts of individuals dedicated to nematological research. However, in the mid-1980's, with the retirement of the "pioneer" nematologists, major changes were taking place. S. Van Gundy commented, "since many of today's nematologists were trained in the early 1950's, our Twenty-fifth Anniversary marks the approaching retirement of the bulk of our charter members. For example, on [SON's] Twenty-fifth Anniversary, seven [of the 14 charter members from the University of California system] have passed away or retired and the other seven are within 5 years of retirement age. This has meant an almost complete turnover in the staffing at the Division of Nematology at the University of California, Davis. Elsewhere, the situation is the same. Thus, this anniversary marks a transition between the older era of nematologists and the new generation of members of our discipline. These new, bright, young nematologists and a new generation of members of other disciplines such as biochemistry, physiology, microbiology, genetics, molecular biology, and computer science...will provide the new direction in nematology."

Nematology positions left vacant by retiring nematologists were being filled by a new generation of scientists who were concentrating their work in the field of biotechnology. Biotechnology and the technological revolution of computers and internet began to dramatically change modern agriculture and in turn, the emphasis of emerging scientists. Farmers began using information technology and precision techniques and by 1994 were using satellite technology to track and plan their farming practices. H. Ferris stated, "the new technology stepped us up to a new level. We started to address all those areas and improved enormously how we grew crops and the productivity of crops."

The new generation of nematologists coming aboard in the 1980's had a different approach to "basic research" as many of the young scientists were being trained in the fields of molecular biology and genetics. V. R. Ferris observed, "when *C. elegans* came along [as a model organism], we were able to approach things from an entirely new point of view. We were learning basic truths about how pests and plants behaved." T. Vrain, SON President (1999-2000), commented, "with the advent of molecular biology, we have seen more fundamental sciences such as molecular genetics and breeding, biochemistry, genetic engineering, genomics and others, come to the forefront with the promise that our understanding of nematode parasitism at the molecular level will lead to an environmentally innocuous management strategy."

Along with the emphasis on biotechnology, biocontrol also became a priority during this period. For decades, nematologists focused on the applied aspects of the science. At the outset of the 1990's, the chemical tools for nematode control were further endangered with discontinuation of fumigants such as methyl bromide, the increased regulations on nematicides which were still available on the open market and the lack of progress being made in developing practical and reliable alternatives to nematicides. D. McClure, retired farm manager for Maui Pineapple Hawaii, stated, "the best nematicides we ever had and probably will ever have are now gone and we are not going to have them back. They were persistent; they lasted; they were fairly toxic; and they worked great on nematodes."

Due to heightened restrictions of chemical pest control, nematologists and chemical companies began to search for a commercial production process for an environmentally friendly nematicide. R. Bolla, SON President (2000-2001), stated that there was "increased interest in using nematodes to control other pests and in developing biological control to manage plant-parasitic nematodes." D. E. Meyerdvink commented, "biological control is increasingly in demand as the 'method of choice' for managing arthropod, disease and weed pests in American agriculture and the environment. The shift in focus from a chemical to a biological basis of Integrated Pest Management is particularly important when using non-indigenous agents in classical biological control. It also requires that biological control helps solve pest problems while improving environmental quality."

Genetic engineering was also implemented as an alternative to chemical controls. "Genetic engineering," C. H. Opperman, molecular nematologist, North Carolina State University, stated, "made it possible to manipulate the most basic components of a living organism to improve its performance in the field without application of costly and potentially hazardous chemicals." For example, in 1994, the Food and Drug Administration (FDA) granted approval of the FLAVR-SAVR tomato, a whole food produced through biotechnology. In 1997, the first herbicide tolerant and the first insect-resistant "biotech" crops for soybeans and cotton were available commercially.

Even with these advances, there were individuals who were hesitant to embrace the technology of molecular biology. Jim Mueller, Dow AgroSciences, remarked that "people fear what they don't understand. It's natural. People fear genetically engineered crops. The European community [was] opposed to them [in 2005]." However, despite some opposition, use of genetically engineered crops were increasing. C. H. Opperman stated that "more people are accepting transgenic approaches...We should take a page from that in nematology. It is a tremendous opportunity to be out on the leading edge and explain what it is going to do."

Research conducted in the areas of biotechnology, biocontrol and the green industries such as landscaping, home horticulture and urban forestry received priority for government funding. With more government money being allocated to other issues, the science of nematology began to feel an economic setback. R. D. Plowman, USDA Administrator, in a letter to R. Rodriguez-Kabana, SON President (1988-1989), wrote that "the number of research nematologists has declined despite the intent, expressed in our 6-Year Implementation Plan, to increase allocations for plant parasitic research on nematodes by 30%. This was due to the requirement to respond to new priorities (food safety, groundwater contamination, etc.) in the absence of increased buying power in our appropriation. Thus, new priorities were met by redirecting resources-impacting many other disciplines along with nematology-that become unencumbered through attrition of scientists. Since 1982, more than $70 million has been shifted to meet new priorities. We do not intend to further weaken nematology research. On the contrary, we recognize the serious losses caused by nematodes and intend to strengthen relevant research programs when such opportunities arise."

For years, money for the support of nematological research had been provided by the state and federal governments (e.g., Hatch Funds). However, following the 1980's, heavy emphasis was placed on obtaining grants to support research. D. Bird recalled, "we are in an era where scientists are expected to attract their own funding. The days of Hatch Funding are gone. My father [Allen Bird] never had to write a grant. If he wanted something, he just went and asked for it. I was brought up in a culture to go out and seek money. Now we are expected to compete for our money." S. A. Lewis commented that "the competition for grants is fierce. The success rate is 12%." If the faculty member was not able to fund the technician with grant funds, then the person would have to be terminated." In 1996, the United States government proposed substantial budget cuts for agricultural research.

In addition to the decreased government funding, potential progress in nematology was endangered by the continued reduction in the number of nematologists. R. Bolla observed, "in the United States, there are only a few specifically defined nematology departments or units. Most academic nematologists are in plant pathology or entomology departments with a shared department name. A similar situation exists in Canada and Brazil. In the US, the largest academic nematology programs are concentrated in the Southeast, Midwest and California where the demand for support of crop production remains very high...The nematology community continues to thrive in Europe in locations where sugar beet and potatoes are produced. These groups are in the United Kingdom, Belgium, Germany and the Netherlands where they remain significantly strong and continue to add nematologists to their units upon retirements. Even in some cases [they] have been able to create new positions. It appears that there is a decline in nematologists involved in identification, systematics and description. This, in part, is a result of the shift towards molecular systematics where there actually might be an increase in research interest. Another part of the decline is a decrease in the number of nematologists involved in extension diagnostic work...It is clear that there is a great demand for nematologists working in molecular biology and the development of new management systems."

Nematology departments and other academic units with nematologists were being reduced in number of positions, merged with other departments or eliminated. In many cases, vacant positions, created by retiring nematologists, were "swept" by university presidents and college deans. R. D. Riggs commented, "in recent years, the number of nematology positions have been declining because funding for agricultural research has not kept up with inflation or the needs. Nematologists who retired or took administrative positions are not being replaced or being replaced with biotechnologists that usually are not interested in research on nematodes. Because of the lack of available jobs, good students are not beating paths to our door to get advanced degrees specializing in nematodes." L. R. Krusberg, SON President (1991-1992), commented, prior to his retirement, "I have been informed that if and when I retire, my position most likely will be filled by a 'molecular' nematologists if it is filled by a nematologists at all."

As a result of downsizing, the general enrollment in colleges of agriculture declined. In 1990, there were less then thirty students pursuing advanced degrees in nematology in the United States and of these, less than fifteen were doctoral students. Also, funding for applied nematology became difficult to procure. H. Ferris stated that at the University of California-Davis, "it is a land grant university. It has been historically a very important College of Agriculture. I teach a course in plant nematology. Over the years, I had up to a dozen students with majors in crop science. There are not any more than a dozen majoring in crop science now. There is a reduction in the interest of undergraduates coming into agriculture. The reduction in flow is exacerbated by economic conditions at our university that impose financial restrictions." For example, in the 1980's, North Carolina State University boasted of 11 nematology graduate students. A few years later, there was only one student studying nematology. J. N. Sasser speculated that students were not too interested in nematology due to "a lack of urgency." He proposed the idea that good nematode control might have reduced the awareness and desire to study management methods.

Experiment stations were also losing personnel during this era and were no longer seen as a source of information for farmers. F. P. Miller stated, "one of the 'costs' of [broadening our scope in agricultural related areas] coupled with our reduced faculty number was that our visibility and service to traditional agricultural constituents have been reduced, thereby giving them the impression that they have been devalued and that we are not as interested in them as we once were." T. Niblack, SON President (2004-2005), commented, "farmers go to their seed man for their information because they have personal relationships with those people." She emphasized that farmers obtain most of their information today from the private sector and much less from the university outreach programs. D. P. Schmitt, SON President (1989-1990), working with a Sociologists from North Carolina State University from 1996 to 1997, conducted a series of focus groups with farmers in North Carolina and Iowa. These growers clearly stated they they could trust the information from the University sources, but still felt that they preferred to go to seed and chemical dealers for their information. Ironically, as stated by T. Niblack, "the dealers get their information from us. With fewer county agents, the academic outreach contact is not as accessible as they once were. In contrast, [farmers] have a seed dealer that is two miles down the road. They have an economic relationship with that person. I think that they rely much more on information that comes from there."

The losses experienced at universities and experiment stations reflected the societal shift away from agriculture. The

1990 Census recorded that most Americans lived in urban areas with less than 2% of the US work force being comprised of farmers. F. P. Miller commented that "agriculture was no longer viewed as an institution or life style to be accorded special protection...Agriculture was being more commonly viewed by the public as a competitor for natural resources." The USDA's Economic Research Service reported that "rural America has changed in many ways over the century. The rural economy in particular has changed-shifting from dependence on farming, forestry, and mining to striking diversity of economic activities...Rural America became home to a smaller and smaller share of the Nation's population. And while it continues to provide most of the Nation's food and fiber, rural America has taken on additional roles, providing labor for industry, land for urban and suburban expansion, sites for storage of waste and hazardous activities, and natural settings for recreation and enjoyment."

By the late 1990's, the era, which witnessed the advances of the "Founding Fathers" of SON, came to a close. A new breeze began to blow as the science of nematology experienced changes in agricultural emphasis, governmental support and technological advances. The new generation of nematologists started shaping the future of the science of nematology by adapting to the emerging need for "earth-friendly" practices and embracing modern technology in the field of molecular biology. As the new millennium dawned, nematologists began to see the impacts of these changes.

*As is obvious to anyone fortunate to be a nematologists, the past decade has seen an explosion of international collaboration among nematologists. Although this collaboration has been necessitated by the multi-disciplinary nature of nematological research, this internationalism has also resulted from improvements in global communication and the expansion of the global economy.*

　　　　　　　　　　-President D. J. Chitwood
　　　　　　　　　　 SON President

# 8 The New Millennium

With the dawning of a new millennium, the world was in a state of uncertainty. As the clocks rolled nearer to 2000, society waited impatiently to see if the new technology would crash and the world would plummet into darkness. Yet, the sun rose on the new millennium without fail. Mirroring the world's anticipation of the new millennium was the world of nematology. Technological advances and emerging scientific disciplines created setbacks for nematology in the 1980's and 1990's. As nematological funding began to disappear and nematology positions diminished, nematologists began to fear for the future of their science.

　　Reaction to the decline in nematology began in the early 1990's through a proposal generated by the SON Long Range Planning Committee. In response, the Society of Nematologists began to explore avenues in which to address key issues effecting their discipline. The Society of Nematologists, collaborating with the Cooperative State Research Service (CSRS) [now known as the Cooperative State Research, Education, and Extension Service (CSREES)], established a committee to develop a perspective and identify research, education and technology-transfer priorities for nematology. This committee, chaired by K. R. Barker, conducted a study of the contributions, status, challenges, needs and goals of plant and soil nematology. The Committee compiled its results into a report entitled *Plant and Soil Nematodes: Societal Impact and Focus for the Future.* It examined soil and plant nematology in an attempt to demonstrate the importance of nematodes and the serious damage they caused to agriculture. The report promoted the lessening of the negative societal impact of plant-parasitic nematodes, including the development of alternative strategies to hazardous pesticides, advancing the knowledge of fundamental nematode biology and promoting the beneficial use of nematodes. It proposed replacing lost nematology positions, offering grants to expand graduate education in nematology and making financial resources available for nematological research. The study was distributed to federal, state and private institutions of higher learning in the United States. Nematologists were encouraged to contact policy makers to enhance their awareness of nematodes and the role they played in agriculture, horticulture and the environment. The report "detailed the current state of US nematology and presented a concrete proposal to improve it," stated E. C. Bernard, JON Editor (1997-2000). The awareness generated by the report, in addition to a general sense of urgency in the discipline of nematology, made it apparent that a major effort was needed to promote the science of nematology. The decision was made to hire a lobbyist to represent SON in Washington, D. C. The issue was debated at a special session during the Thirty-third Annual Meeting of Society of Nematologists at San Antonio, Texas in 1994 and agreed upon. However, the lobbyist was never hired and efforts to enhance the image of nematology began to diminish throughout the remainder of the twentieth century.

Besides the issues being addressed in regard to the decline in nematology in the United States, international restrictions placed on agricultural products worldwide as well as an ever increasing need to foster global cooperation became matters of urgency. H. Ferris stated, "we regulate the use of pesticides in the production in our state from the stand point of food stuff and environmental health. We cannot ignore the production of safe foods in other countries. If we are doing it [growing crops] globally, and we live in a global environment, we need to be training and reaching out and working on a more international stage."

In an attempt to embrace international interaction, the idea of inter-societal cooperation surfaced as a means to revitalize the science of nematology worldwide. K. Barker, first President of the International Federation of Nematology Societies (IFNS) (1996-2002), stated, "the current recession and downsizing of nematology clearly poses challenges to our small discipline and threatens its survival. This situation could catalyze nematologists to broaden our horizons, including international and local scientific public and industry linkage."

In 1990, at the Second International Nematology Congress in The Netherlands, S. A. Lewis, SON President (1990-1991), established a committee of representatives from eight societies, chaired by K. R. Barker, to examine the feasibility of an international federation. Lewis stated, "if the confederation can promote nematology, perhaps by providing an umbrella for the formation of nematology societies in different countries, by sponsoring international meetings and by publishing a newsletter, a case can be made for the formation of such a group."

The aim of the International Federation of Nematology Societies (IFNS) was to hold international congresses on nematology in which to provide a forum to foster communication among nematologists worldwide in an effort to enhance awareness of nematodes and the science of nematology. J. L. Starr, SON President (1996-1997), commented, "the International Federation took twelve years to organize. Every time we had a meeting to discuss it, we had all of these people who came, talked about it from different societies but they were not empowered to act. They went back to discuss it and there would be another point raised. We would meet some place the next year and discuss it again and this point would be resolved and something else would come up but no one was again empowered to make a decision...It went on forever." K. R. Barker recalled, "I stored 14,000 emails concerning the debate over the formation of the IFNS."

After years of discussion, IFNS was finally established in July 1996 at the Third International Nematology Congress in Guadeloupe. The nematological societies that formed this umbrella federation included: the Brazilian Nematological Society, Egyptian Society for Agricultural Nematology, European Society of Nematologists, Italian Society of Nematology, Japanese Nematology Society, Nematological Society of Southern Africa, Organization of Nematologists of Tropical America, Pakistan Society of Nematology, the Russian Society of Nematology and the Society of Nematologists.

In addition to the formation of IFNS, emphasis was placed on promoting interaction between different nematological societies in order to enhance the visibility of nematology. During this period, the membership of nematological societies was declining. For example, SON membership dropped from 915 members in the mid-1980's to 650 members by the mid-1990's (Figure 8-1). D. Chitwood, SON President (1995-1996), observed that "nematology has grown far more complex and interdisciplinary. As research and extension activities have broadened, many nematologists have found that their professional careers benefit from attendance at a diverse range of professional meetings." Due to this trend, scientific societies began to realize the need to work together for the advancement of nematology. Nematological organizations

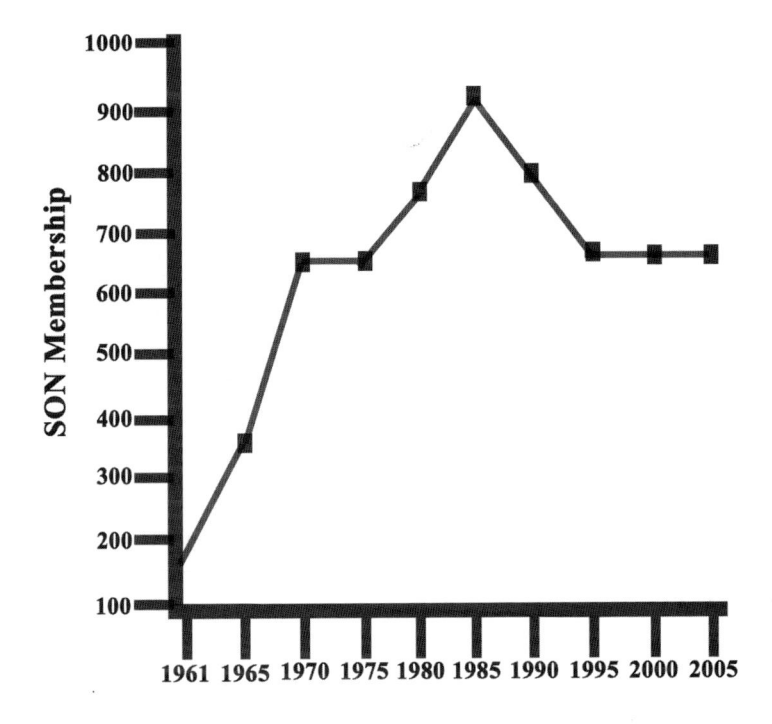

Figure 8-1. Annual SON Membership from the founding of the Society in 1961 to 2005.

that worked independently during the 1990's to 2000's lacked the financial resources to influence public opinion or the legislative process. E. C. Bernard, SON President (1992-1993), added that "by speaking with collective voices, we can be more effective in maintaining the high level of visibility we need." D. Bird commented that "in the 1960's to 1980's, we saw plant nematology become a very separate discipline. It was a more vibrant discipline with more people involved. As we got a bit smaller, the options of more interaction presented themselves. By thinking broadly across disciplines we can understand there are some underlying principles."

One such collective effort was the formation of the

Coalition on Funding Agricultural Research Missions (CoFARM) in 1990. This group was comprised of 16 scientific societies, with a combined membership of 100,000. The goals of CoFARM were to effectively educate policy makers about the value and importance of public investments in agricultural research and to focus the USDA research budget on agricultural investments that would enhance the quality of life of Americans and increase US competitiveness in expanding world markets.

The global markets began to play an important role in agricultural production during the 1990's. Every facet of agricultural production began to shift towards incorporating environmentally "friendly" techniques in the production of "safe" agricultural products that would be accepted in the world market place. Concerning the marketing of chemical products, J. P. Mueller stressed that a potential product must have global sales appeal to justify its development. He said, "if we don't have a global product, its very unlikely that we will use limited company resources to launch it." He further explained that the European Union has become an important market for chemical company products. "Generally, if you are developing a new pesticide, you would like the European market. There is a need for global acceptance. The regulatory community is kind of heading in that direction. They talk about harmonization. The US EPA's review is harmonized with the EU's review. They are working on it from that standpoint. It is a global community and the products need to be available in those areas. The products treated with those chemicals here are exported. The export markets are huge for us. They need to be compatible. There is a lot more awareness now."

In an effort to increase the visibility of nematology, the Society of Nematologists initiated a landmark venture in the late 1990's, the founding of the Nathan A. Cobb Nematology Foundation. The mission of this foundation was to build self-

**Figure 8-2**. First N. A. Cobb Nematology Foundation Board. Left to Right: L. W. Duncan, Chair; J. L. Starr, Vice Chair; P. King, Secretary and D. P. Schmitt, Treasurer.

sustaining endowments for the advancement of nematology. L. W. Duncan, N. A. Cobb Foundation Chair (2000-2004), recalled, "I got a phone call...asking me if I would be part of a committee to prepare a constitution for the Cobb Foundation and explore the possibilities. Within two years, we decided that it was feasible and I had drafted the bylaws." The Foundation was incorporated in 1999 in the State of Kansas with its First Executive Board being comprised of L. W. Duncan, Chair, J. L. Starr, Vice Chair, P. King, Secretary and D. P. Schmitt, Treasurer.

The Foundation grew quickly and the initial endowments were established. These endowments included: the Economic Nematology Endowment, the Entomophilic Nematology Endowment, the General Fund (now the N. A. Cobb Fund), the K. R. Barker-IFNS Endowment, the Presidents' Endowment and the John M. Webster Outstanding Student Award. L. Duncan commented, "when we went to the First Board Meeting in Quebec City, Canada, Jim Starr proposed and already initiated the Entomophilic Project. It took off like a rocket. He initiated the President's Foundation. Jim has always had a close connection with [the American Phytopathological Society] and this has made him see how

**Figure 8-3.** J. G. Baldwin at the Marine Biology Session at the Forty-forth Annual Meeting in Fort Lauderdale, Florida, 2005.

these things should go. I cannot believe how generous all the [past] presidents have been. Many of the presidents donated a minimum of $1000 to the Presidents' Endowment." Each endowment was designed to provide funds for activities such as travel grants, scholarships, speaker fees, special publications, resource exchanges and public education.

The Foundation goal of $100,000 by 2005 was proposed and realized. To launch the Foundation, SON provided $5000 in start up funds. The Society also provided incentives for investors by committing $5000 in matching funds for each endowment that raised $5000 or more. Within three years of the establishment of the Nathan A. Cobb Nematology Foundation, enough funds were generated to give two awards.

The success of the Nathan A. Cobb Nematology Foundation at the outset of the new millennium was one of the positive enhancements to the science of nematology. Biotechnology and research on *C. elegans* continued to make

The New Millennium 81

**Figure 8-4.** P. Sundararaj, nematologist, University of Idaho, at the Poster Session at Forty-forth Annual Meeting in Fort Lauderdale, Florida, 2005.

advances and gain worldwide recognition early in the new millennium. In 2002, Sydney Brenner, H. Robert Horvitz and John Sulston were awarded the Nobel Prize for medicine for their work on the genetics of organ development and programmed cell death in *C. elegans*. Also, the 2006 Nobel Prize in Physiology or Medicine was awarded to Andrew Fire and Craig C. Mello, for their discovery of RNA interference in *C. elegans*.

As biotechnology was gaining notoriety through ground breaking discoveries and scientific research, some individuals began to use these scientific advancements in the form of bioterrorism. The threats created by bioterrorism became a global concern after the terrorist attack on the World Trade Center in New York, New York on September 11, 2001. This attack permanently changed societies worldwide. Due to potential bioterrorist attacks, scientific societies and the United States Department of Agriculture (USDA) began to foster awareness of the vulnerability of agriculture and plant resources to biological attacks by establishing the National Center for Plant Biosecurity (NCPB). In 2004, President George W. Bush issued the Homeland Security Presidential Directive #9 (HSPD-9) that established a national policy to defend United States agriculture and food supply from terrorist attacks. SON's president A. P. Nyczepir (2001-2002) responded after September 11, 2001 by establishing "an ad hoc committee to deal with bioterrorism as it pertains to nematology. We the SON EB had some discussion about this issue and some of the committee members were opposed to it. However, the majority felt we should pursue it. We did not call it the 'Bioterrorism Ad Hoc Committee' because we thought that it was too strong. We termed it the 'Nematodes and Society Ad Hoc Committee.' We were going to look at nematodes that might have a negative impact on society and then take this information to the appropriate authorities."

The new millennium has brought many changes to the field of nematology. Perhaps the most substantial change has taken place in the scope of the science as it has reached out to join hands with other disciplines and other countries. As S. Van Gundy so wisely prophesied, "nematology has moved into a new era. These new scientists coming into the field must be able to pursue broader research goals and broader visions."

*Whoever wishes to foresee the future must consult the past; for human events ever resemble those of preceding times. This arises from the fact that they are produced by men who ever have been, and ever shall be, animated by the same passions, and thus they necessarily have the same results.*

                         -Machiavelli

# 9 Nematological Horizons

Machiavelli once stated, "whoever wishes to foresee the future, must consult the past." As SON approaches its Golden Anniversary, it is important to look back at the Society's interaction with modern agriculture, the discipline of nematology and the Society of Nematologists itself.

Almost sixty years have past from the agricultural boom in the 1950's to the technological advances witnessed at the start of the new millennium. Throughout this time period, agriculture progressed from small family farms to corporate farming. Farmers witnessed the advancement from the horse pulled plows to modern farming techniques which incorporated genetically modified organisms (GMO) and satellite tracking systems. These years also witnessed the introduction of ground breaking chemical pest controls and its widespread elimination due to environmental concerns.

Similarly, the discipline of nematology has experienced dramatic changes over the past 60 years. Nematology, once a relatively unknown science, exploded in the 1950's to the 1970's with the introduction of chemical nematicides. Awareness of declining crop yields caused by nematodes increased government and private funding. These funds enabled nematologists to conduct research, enhancing their knowledge of nematodes, allowing them to embrace new technologies and helping them to control nematodes. Within a few decades, these technologies enabled this specialized field of study to change dramatically. As G. Thorne commented, much has changed from the days when individuals "recommended placing a spoonful of soil on a piece of 'scottie' in a funnel allowing a few nemas to wiggle out of it," to the present day use of biotechnology. K. R. Barker observed that "the changes in nematology during the past 50 years, from using mechanical calculators and typewriters to high-capacity computers for data management, development of nematode-host-environment simulation, basic physiological work to molecular biology and related analytical equipment, including development of DNA probes, cloning genes, engineering host resistance and genomics, ecology-based nematode-IPM-crop management to using the Internet to instantly sharing information worldwide- is mind boggling."

Paralleling the changes and growth in both agriculture and nematology, the Society of Nematologists has evolved greatly during its brief but rich history. In the late 1950's and early 1960's, dedicated nematologists conceptualized and brought into fruition a society to foster the growth of a small emerging discipline. The seeds of SON were sown by these members as well as nurtured to maturity during the first three decades of the Society's existence. The care of this organization was passed on to a new generation who, once entrusted with its future, embraced the new technologies, molecular biology and the environmental practices of their era in order to enhance and further the science of nematology.

Although changes have occurred in agriculture, nematology and the Society of Nematologists, there are certain truths that will never change. Agriculture has always and will always remain a necessity to sustain and perpetuate humankind. Nematodes will continue to create challenges as well as offer possible benefits to individuals working in the discipline of nematology. Scientists will continue to assist agricultural workers by conducting research on these organisms, using whatever means and technology they deem best for the advancement of their science. The Society of Nematologists will continue to offer a forum whereby all who are dedicated to this specialized field can work together for the advancement of the science of nematology.

The history of agriculture, nematology and the Society of Nematologists has filled numerous chapters to date, leaving many pages yet to be written. What will those blank pages one day read? According to B. Endo, SON President (1972-1973), "the future of nematology is wide open and this is the main thing that needs to be known." As this volume comes to a close, the torch is being passed to those coming after, to carry on the traditions established by the founding fathers of SON and nematology. Who will pen the next volume? What will fill its pages? The future characters and events are yet to be determined. The pages of the future wait to be written.

## References

ADAMS, B. 2007. NNL interview with Dr. Vernon G. Perry, president of SON 1966-1967. Nematology Newsletter 52(3-4):8.

ALLEN, M. W. 1958. Nematodes as parasites of plants. Pp. 2-4 *in* Proceedings of the Shell Nematology Workshop, February 12-13, 1958, Yuma, Arizona.

ALLEN, M. W. 1962. Personal communication. February 1, 1962.

AYCOCK, R. 2005. Personal communication. January 26, 2005.

AYERS, E. L. 1957. The importance of the nematode problem in the state of Florida. P. 59 *in* Proceedings of the Shell Nematology Workshop, August 21, 1957, Orland, Florida.

BALDWIN, J. 1998. From the president. Nematology Newsletter 45(2):1.

BALDWIN, J. 2005. Personal communication. July 11, 2005.

BARKER, K. R. 1992. International federation of nematology societies established. Nematology Newsletter 42(3):9.

BARKER, K. R., R. S. HUSSEY, L. R. KRUSBERG, G. W. BIRD, R. A. DUNN, H. FERRIS, D. W. FRECKMAN, C. J. GABRIEL, P. S. GREWAL, A. E. MACGUIDWIN, D. L. RIDDLE, P. A. ROBERTS AND D. P. SCHMITT. 1994. Plant and soil nematodes: societal impact and focus for the future. Journal of Nematology 26(2):127-137.

BARKER, K. R. 1998. Nematology: then, then and now. Nematology Newsletter 44 (1):24-27.

BARKER, K. R. 2003. Perspectives on plant and soil nematology. Annual Review of Phytopathology 41:1-25.

BARKER, K. R. 2004. Personal communication. August 10, 2004.

BARKER, K. R. 2005. Personal communication. January 27, 2005.

BERNARD, E. C. 1993. From the president. Nematology Newsletter 39(3):1.

BIRD, D. M. 2004. High society (of nematologists). Genome Biology 2004(5):353.

BIRD, D. M. 2007. Personal communication. April 5, 2007.

BOLLA, R. 2000a. From the president. Nematology Newsletter 46(3):1.

BOLLA, R. 2000b. From the president. Nematology Newsletter 46(4):3-4.

BOUSKA, E. 2005. Personal communication. August 19, 2005.

CASWELL-CHEN, E. 2004. Personal communication. August 9, 2004.

CHITWOOD, D. W. 1996a. From the president. Nematology Newsletter 42(1):1.

CHITWOOD, D. W. 1996b. From the president. Nematology Newsletter 42(2):1.

CHITWOOD, D. W. 1996c. Minutes from the first executive board meeting. Nematology Newsletter 42(4):10.

CHRISTIE, J. R. 1951. Personal communication. June 15, 1951.
CHRISTIE, J. R. 1957. Personal communication. May 15, 1957.
COOK, M. G. 1990. Soil Facts LISA: current status and future outlook. <http://www.soil.ncsu.edu/publications/Soilfacts/AG-439-07> April 4, 2007.
DOUCET, M. 1984. Personal communication. December 14, 1984.
DROPKIN, V. H. 1962. Personal communication. April 9, 1962.
DUNCAN, L. W. and D. P. SCHMITT. 2002. Building a foundation. Nematology Newsletter 48(1)3.
DUNN, R. A. 1998. The ups and downs of extension nematology. Nematology Newsletter 44(1):19.
ECONOMIC RESEARCH SERVICE. 2005. Understanding rural America. <http://www.ers.usda.gov/publications/aib710/index.htm> June 7, 2005.
EISENBACK, J. D. 1997. Minutes of executive board and business meetings. Nematology Newsletter 43(3):7.
ENDO, B. 2004. Personal communication. August 10, 2004.
FERRIS, H. 2005. Personal communication. July 12, 2005.
FERRIS, J. M. 1964. Personal communication. May 9, 1964.
FERRIS, V. R. 1964. Personal communication. March 30, 1964.
FERRIS, V. R. 1968. Personal communication. March 11, 1968.
FERRIS, V. R. 1977. Nematology. Communicator: 3(2):3-5.
FERRIS, V. R. 2005. Personal communication. July 12, 2005.
FOREST, A. A. 1957. The importance of nematodes in forestry. P. 28 *in* Proceedings of the Shell Nematology Workshop, December 18-19, 1957, Columbia, South Carolina.
GOLDEN, A. M. 1967. Initation of a journal. Nematology Newsletter 13(3):3.
GOLDEN, A. M. 1969. Personal communication. January 31, 1969.
GOOD, J. M. 1957. The role of plant parasitic nematodes in peanut production. P. 15 *in* Proceedings of the Shell Nematology Workshop. December 18-19, 1957, Columbia, South Carolina.
GOOD, J. M. 1965. Nematology investigation. Crops Protection Research Branch, Crops Research Division, Agricultural Research Service, US Department of Agriculture, Beltsville, Maryland. Nematology Newsletter 11(2):8.
HART, W. H. 1958. Regulatory aspects of plant nematology. P. 21 *in* Proceedings of the Shell Nematology Workshop, February 12, 1958, Yuma, Arizona.
HEALD, C. M. 1987. Classic nematode management practices. Pp. 100-104 *in* J.A. Veech and D. W. Dickson, editors. Vistas on Nematology. Maryland: Society of Nematologists.
HOLDEMAN, Q. L. 1956. A new regional project organized, NE-34.

Nematology Newsletter 1(2):1-2.
HOLDEMAN, Q .L. 1957. NC-28 organized. Nematology Newsletter 2(1):1.
HOLDEMAN, Q. L. 1958a. Holdeman bids farewell. Nematology Newsletter Vol. 3(4):1.
HOLDEMAN, Q. L. 1958b. Many nematologists attend AIBS meetings. Nematology Newsletter 3(3):3.
HOLDEMAN, Q. L. 1958c. Meetings. Nematology Newsletter 3(2):3.
HOLDEMAN, Q. L. 1958d. Shell nematology workshops. Nematology Newsletter 3(1):1.
HOLDEMAN, Q. L. 1958e. Summer session in nematology - 1959. Nematology Newsletter 3(2):1.
INTERNATIONAL FEDERATION OF NEMATOLOGY SOCIETIES. 1992. Draft: upadate on continued study on feasibiity of proposed International Federation of Nematology Societies. July 1, 1992.
INTERNATIONAL FEDERATION OF NEMATOLOGY SOCIETIES. 2005. A global communication forum <http://www.ifns.org/about.html> February 1, 2005.
INTERSOCIETY CONSORTIUM FOR PLANT PROTECTION. 1983. Intersociety consortium for plant protection. St. Paul: ICPP.
JENSEN, H. J. 1959. The scope of nematodes as plant parasites. P. 7 *in* Proceedings of the Shell Nematology Workshop, January 27-28, 1959, Portland, Oregon.
JENSEN, W. R. 1963. Personal communication. March 12, 1963.
JENSEN, W. R. 1969. Emblem Design. Nematology Newsletter 15(4):3.
KING, I. 2005. Personal communication. July 11, 2005.
KRUSBERG, L. 1992a. From the president. Nematology Newsletter 38(1):1.
KRUSBERG, L. 1992b. From the president. Nematology Newsletter 38(2):1.
KRUSBERG, L. 1993. Annual reports. Nematology Newsletter 39(3):8.
LEWIS, S. A. 1990a. Personal communication. July 25, 1990.
LEWIS, S. A. 1990b. From the president. Nematology Newsletter 36(4):1.
LEWIS, S. A. 2005. Personal communication. July 10, 2005.
LISTAIR, J. and B. HYMAN. 2002. The continued evolution of the Journal of Nematology. Nematology Newsletter 48(4):13.
MAI, W. F. 1969. Personal communication. August 9, 1969.
MAI, W. F. and R. E. MOTSINGER. 1987. History of the society of nematologists. Pp. 1-6 *in* J.A. Veech and D. W. Dickson, editors. Vistas on Nematology. Maryland: Society of Nematologists.
MCCLURE, M. A. 1998. From the president. Nematology Newsletter 44(1):1.
MILLER, F. P. 1993. Forces and factors driving changes in colleges of agriculture: the dissolution of the department of agronomy at the Ohio

State University. COFarm 10/13 p. 1-6.
MILLER, L. 1968. Personal communication. October 4, 1968.
MORTON, H. V. 1987. Industry perspectives in nematology. Pp. 47-51 Pp. 1-6 in J. A. Veech and D. W. Dickson, editors. Vistas on Nematology. Maryland: Society of Nematologists.
MOUNTAIN, W. B. 1963. Personal communication. September 17, 1963.
MOUNTAIN, W. B. 1966. Personal communication. October 3, 1966.
MOUNTAIN, W. B. 1967. The Society of Nematologists - where should it go from here? Nematology Newsletter 11(2):3.
MEYERDVINK, D. E. 1995. Personal communication. April 18, 1995.
NGUYEN, K. B. 2003. History of nematology with photographs of nematologists. <http://www.flnema.ifas.ufl.edu/history/nem_history.htm> January 6, 2005.
NORTON, D. Personal communication. July 20, 2004.
NYCZEPIR, A. 2005. Personal communication. February 10, 2005.
OPPERMAN, C. H. 1997. Editorial. Nematology Newsletter 43(4):4.
OPPERMAN, C. H. 2007. Personal communication. April 10, 2007.
PERRY, V. G. 1958. History of nematode control. Pp. 1-3 in Proceedings of the Shell Nematology Workshop, February 20-21, 1958, Toldeo, Ohio.
PLOWMAN, R. D. 1988. Personal communication. September 23, 1988.
RACHEL CARSON.ORG. 2007. Biography of Rachel Loise Carson. <http://www.rachelcarson.org/index.cfm?fuseaction=bio> April 2, 2007.
RASKI, D. J. 1959. Personal communication. April 2, 1959.
RASKI, D. J. 1959. Personal communication. January 6, 1959.
RIGGS, R. D. 1995. Long range planning committee: planning for the future of the Society of Nematologists. Nematology Newsletter 41(2):4.
RIGGS, R. D. 2004. Personal communication. August 8, 2004.
SASSER, J. N. 1959. Personal communication. March 19, 1959.
SASSER, J. N. and W. R. JENKINS. 1960. Nematology: fundamentals and recent advances with emphasis on plant parasitic and soil forms. Pp. 1-2. Chapel Hill: The University of North Carolina Press.
SASSER, J. N. 1963a. Personal communication. April 16, 1963.
SASSER, J. N. 1963b. Personal communication. October 25, 1963.
SASSER, J. N. 1964. Nematology and the role of the Society of Nematologists in the advancement of this science. Presidential Address. P. 4. Annual Meeting, Society of Nematologists, University of Colorado, Boulder, Colorado, August 23-28, 1964.
SASSER, J. N. 2005. Personal communication. January 26, 2005.
SCHMITT, D. P. 2007. Personal communication. April 15, 2007.
SEINHORST, J. W. 1957. Phytonematology in western Europe. Auburn, AL:

Southern Regional Nematology Project.
SHEET, T. J. 1983. A proposal to hold a workshop concerned with pest management in transition - the role of the EPA. St. Paul: ICPP.
SIPES, B. 2003. History of Nematology. Nematology Newsletter 49(3):21.
SLACK, D. A. 1959a. Proceedings of the S-19 workshop in phytonematology. Nematology Newsletter 5(1):2.
SLACK, D. A. 1959b. Teaching in nematology. Nematology Newsletter 5(2):1.
SLACK, D. A. 1959c. Society of nematologists? Nematology Newsletter. Vol. 5(3):1-3.
SLACK, D. A. 1959d. Society of nematologists? Nematology Newsletter. Vol. 5(4):2.
SLACK, D. A. 1961a. American Nematological Society. Nematology Newsletter 7(2):2.
SLACK, D. A. 1961b. American Nematological Society. Nematology Newsletter 7(3):3.
SLACK, D. A. 1962. Report of the 1962 annual meeting of the Society of Nematologists. Nematology Newsletter 8(3):1.
SLACK, D. A. 1963. Personal communication. February 27, 1963.
SMART, G. C. 1967a. Editoral comments. Nematology Newsletter 13(3):3.
SMART, G. C. 1967b. On beginning a journal. Nematology Newsletter 13(1):3-4.
SOCIETY OF NEMATOLOGISTS EDITORIAL BOARD MEETING. 1967. Beltsville, MD: November 1, 1967.
SPEARS, J. F. 1968. History and orgin of the golden nematode. <http://www.ceris.purdue.edu/napis/pests/gn/handbook.html> March 12, 2007.
STARR, J. L. 1996. From the President. Nematology Newsletter 42(4):1.
STARR, J. L AND B. HYMAN. 2002. The continued evolution. Nematology Newsletter 48(4):14.
STARR, J. L. 2005. Personal communication. July 10, 2005.
STAYNER, L. F. 1958. Welcome and definition. P. 1 in Proceedings of the Shell Nematology Workshop, February 12, 1958, Yuma, Arizona.
STURGIS, J. T. 1958. Review of experiments with the commercial application of soil fumigants. Pp. 1-3 in Proceedings of the Shell Nematology Workshop, Feburary 20-21, 1958, Toledo, Ohio.
TARJAN, A. C. 2004. Personal communication. August 9, 2004.
TAYLOR, A. L. 1978. Nematocides and nematicides - history. <http://flnema.ifas.ufl.edu/history/nematocide_his.htm.> January 6, 2005.
TAYLOR, D. P. 1959a. Personal communication. August 22, 1959.
TAYLOR, D. P. 1959b. Personal communication. September 8, 1959.

TAYLOR, D. P. 1961. Minutes of the first meeting of the executive committee of the Society of Nematologists.
TAYLOR, D. P. 1962. Personal communication. March 14, 1962.
THE INDUSTRIAL REVOLUTION. <http://www.bergen.org/technology/indust.html> July 4, 2006.
THORNE, GERALD. 1957. Plant parasitic nematodes in soil biology. Pp. 1-2 in Soil Science Society of America. Cincinnati: Soil Science of America.
THORNE, GERALD. 1959. History of nematode control. Pp. 1-2 in Proceedings of the Shell Nematology Workshop. Portland: Shell Nematology Workshop.
THORNE, GERALD. 1961. Principles of Nematology. New York: McGraw Hill.
VAN GUNDY, S. D. 1968. Personal communication. March 5, 1968.
VAN GUNDY, S. D. 1980. Nematology-status and propspects: lets take off the blinders and broaden our horizons. Journal of Nematology 12(3):158-163.
VAN GUNDY, S. D. 2004. Personal communication. August 9, 2004.
VRAIN, T. 2001. From the president. Nematology Newsletter 46(1):1.
WILSON, C. T. 1990. Development of an Expanded Program of Teaching and Research in Nematology in the Southern Region in Southern Cooperative Series Bulletin 276.
UNITED STATES DEPARTMENT OF AGRICULTURE. 2005a. A history of American agriculture: agricultural education & extension. <http://www.agclassroom.org/gan/timeline/ag_edu.htm> June 7, 2005.
UNITED STATES DEPARTMENT OF AGRICULTURE. 2005b. A history of American agriculture: farmers & the land. <http://www.agclassroom.org/gan/timeline/ag_trade.htm> June 7, 2005.
UNITED STATES DEPARTMENT OF AGRICULTURE. 2005c. A history of American agriculture:life on the farm. <http://www.agclassroom.org/gan/timeline/ag_edu.htm> June 7, 2005.
UNITED STATES ENVIRONMENTAL PROTECTION AGENCY. 2007. Introduction to clean water act. <http://www.epa.gov/watertrain/cwa> April 2, 2007.
VAN GUNDY, S. 1969. Dedication. Journal of Nematology 1(1):1.
VAN GUNDY, S. 1987. Prespectives on nematology research. Pp. 28-31 in J. A. Veech and D. W. Dickson, editors. Vistas on Nematology. Hyattsville, MD: Society of Nematologists.
WALL, D. H. 2005. Personal communication. 15 February 2005.
WEBSTER, J. M. 1980. Nematodes in an overcrowded world. Revue de Nematologica 3(1):135-143.
WEBSTER, J. M. 1984. Personal communication. November 21, 1984.

WEBSTER, J. M. 1998. Nematology: from curitostiy to space science in fifty years. Annals of Applied Biology 132:3-11.

WIKIPEDIA. 2007. *Caenorhabditis elegans.* <http://en.wikipedia.org/wiki/C. elegans> April 5, 2007.

# Appendix I
## Society of Nematologists Presidents

M. W. Allen
1961-1963

J. N. Sasser
1963-1964

W. B. Mountain
1964-1965

V. H. Dropkin
1965-1966

V. G. Perry
1966-1967

A. M. Golden
1967-1968

W. F. Mai
1968-1969

V. R. Ferris
1969-1970

J. M. Good
1970-1971

H. J. Jensen
1971-1972

B. Y. Endo
1972-1973

S. D. Van Gundy
1973-1974

J. H. O'Bannon
1974-1975

I. J. Thomason
1975-1976

L. I. Miller
1976-1977

G. W. Bird
1977-1978

# 96  Agriculture, Nematology and SON

C. M. Heald
1978-1979

K. R. Barker
1979-1980

J. M. Ferris
1980-1981

D. W. Dickson
1981-1982

J. M. Webster
1982-1983

D. W. Wall
1983-1984

G. D. Griffin
1984-1985

G. Fassuliotis
1985-1986

J. A. Veech
1986-1987

A. W. Johnson
1987-1988

R. Rodriguez-Kabana
1988-1989

D. P. Schmitt
1989-1990

S. A. Lewis
1990-1991

L. R. Krusberg
1991-1992

E. C. Bernard
1992-1993

R. D. Riggs
1993-1994

# SON Presidents 97

G. S. Santo
1994-1995

D. J. Chitwood
1995-1996

J. L. Starr
1996-1997

M. A. McClure
1997-1998

J. G. Baldwin
1998-1999

T. C. Vrain
1999-2000

R. I. Bolla
2000-2001

A. P. Nyczepir
2001-2002

T. O. Powers
2002-2003

R. Giblin-Davis
2003-2004

T. Niblack
2004-2005

S. Hafez
2005-2006

A. E. MacGuidwin
2006-2007

E. Davis
2007-2008

R. Huettel
2008-2009

R. Ingham
2009-2010

# Appendix II
## Society of Nematologists Officers

| Year | President | President Elect | Vice President | Treasurer | Secretary |
|---|---|---|---|---|---|
| 1961-63 | M.W. Allen | — | J.N. Sasser | V.H. Dropkin | D.P. Taylor |
| 1963-64 | J.N. Sasser | — | W.B. Mountain | A.M. Golden | D.P. Taylor |
| 1964-65 | W.B. Mountain | — | V.H. Dropkin | A.M. Golden | D.P. Taylor |
| 1965-66 | V.H. Dropkin | — | D.P. Taylor | A.M. Golden | V.R. Ferris |
| 1966-67 | V.G. Perry | — | A.M. Golden | E.J. Cairns | V.R. Ferris |
| 1967-68 | A.M. Golden | — | W.F. Mai | E.J. Cairns | V.R. Ferris |
| 1968-69 | W.F. Mai | — | V.R. Ferris | J.H. O'Bannon | B.Y. Endo |
| 1969-70 | V.R. Ferris | — | J.M. Good | J.H. O'Bannon | B.Y. Endo |
| 1970-71 | J.M. Good | — | H.J. Jensen | J.H. O'Bannon | B.Y. Endo |
| 1971-72 | H.J. Jensen | — | B.Y. Endo | W.R. Nickle | G.W. Bird |
| 1972-73 | B.Y. Endo | — | S.D. Van Gundy | W.R. Nickle | G.W. Bird |
| 1973-74 | S.D. Van Gundy | — | J.H. O'Bannon | W.R. Nickle | G.W. Bird |
| 1974-75 | J.H. O'Bannon | — | I.J. Thomason | J.M. Ferris | C.M. Heald |
| 1975-76 | I.J. Thomason | — | L.I. Miller | J.M. Ferris | C.M. Heald |
| 1976-77 | L.I. Miller | — | G.W. Bird | J.M. Ferris | C.M. Heald |
| 1977-78 | G.W. Bird | — | C.M. Heald | D.C. Norton | D.W. Dickson |
| 1978-79 | C.M. Heald | — | K.R. Barker | D.C. Norton | D.W. Dickson |
| 1979-80 | K.R. Barker | — | J.M. Ferris | D.C. Norton | D.W. Dickson |
| 1980-81 | J.M. Ferris | — | D.W. Dickson | R.S. Hussey | D.W. Freckman |
| 1981-82 | D.W. Dickson | — | J.M. Webster | R.S. Hussey | D.W. Freckman |
| 1982-83 | J.M. Webster | — | D.W. Freckman | R.S. Hussey | A.W. Johnson |
| 1983-84 | D.W. Freckman | G.D. Griffin | G. Fassuliotis | D.P. Schmitt | A.W. Johnson |
| 1984-85 | G.D. Griffin | G. Fassuliotis | J.A. Veech | D.P. Schmitt | A.W. Johnson |
| 1985-86 | G. Fassuliotis | J.A. Veech | A.W. Johnson | D.P. Schmitt | S.A. Lewis |
| 1986-87 | J.A. Veech | A.W. Johnson | Rodriguez-Kábana | R.A. Kinloch | S.A. Lewis |
| 1987-88 | A.W. Johnson | Rodriguez-Kábana | D.P. Schmitt | R.A. Kinloch | S.A. Lewis |
| 1988-89 | Rodriguez-Kábana | D.P. Schmitt | S.A. Lewis | R.A. Kinloch | J.G. Baldwin |
| 1989-90 | D.P. Schmitt | S.A. Lewis | L.R. Krusberg | L.W. Duncan | J.G. Baldwin |
| 1990-91 | S.A. Lewis | L.R. Krusberg | E.C. Bernard | L.W. Duncan | J.G. Baldwin |
| 1991-92 | L.R. Krusberg | E.C. Bernard | R.D. Riggs | L.W. Duncan | R.N. Huettel |
| 1992-93 | E.C. Bernard | R.D. Riggs | G.S. Santo | M.A. McClure | R.N. Huettel |
| 1993-94 | R.D. Riggs | G.S. Santo | D.J. Chitwood | M.A. McClure | R.N. Huettel |
| 1994-95 | G.S. Santo | D.J. Chitwood | J.L. Starr | M.A. McClure | J.D. Eisenback |
| 1995-96 | D.J. Chitwood | J.L. Starr | M.A. McClure | J.A. Thies | J.D. Eisenback |
| 1996-97 | J.L. Starr | M.A. McClure | J.G. Baldwin | J.A. Thies | J.D. Eisenback |
| 1997-98 | M.A. McClure | J.G. Baldwin | T.C. Vrain | J.A. Thies | A. MacGuidwin |
| 1998-99 | J.G. Baldwin | T.C. Vrain | R.I. Bolla | T.L. Niblack | A. MacGuidwin |
| 1999-00 | T.C. Vrain | R.I. Bolla | A. P. Nyczepir | T.L. Niblack | A. MacGuidwin |
| 2000-01 | R. I. Bolla | A.P. Nyczepir | T.O. Powers | T.L. Niblack | F. Robinson |
| 2001-02 | A. P. Nyczepir | T.O. Powers | R. Giblin-Davis | P. S. King | F. Robinson |
| 2002-03 | T.O. Powers | R. Giblin-Davis | T. Niblack | P. S. King | F. Robinson |
| 2003-04 | R. Giblin-Davis | T. Niblack | S. Hafez | P. S. King | B. Sipes |
| 2004-05 | T. Niblack | S. Hafez | A. MacGuidwin | R. Ingham | B. Sipes |
| 2005-06 | S. Hafez | A. MacGuidwin | E. Davis | R. Ingham | B. Sipes |

## 100 Agriculture, Nematology and SON

| | | | | | |
|---|---|---|---|---|---|
| 2006-07 | A.. MacGuidwin | E. Davis | R. Huettel | R. Ingham | S. Thomas |
| 2007-08 | E.. Davis | R. Huettel | R. Ingham | B. Sipes | S. Thomas |

# Appendix III
## Journal of Nematology Editor-in-Chiefs

S. D. Van Gundy
1969-1971

G. C. Smart, Jr.
1972-1974

K. R. Barker
1975-1977

B. F. Lownsbery
1978-1980

J. Veech
1981-1983

L. R. Krusberg
1984-1986

R. D. Riggs
1987-1990

D. J. Chitwood
1991-1993

D. W. Dickson
1994-1996

E. C. Bernard
1997-1999

B. Hyman
2000-2002

J. L. Starr
2003-2005

D. Bird
2006-2008

# Appendix IV
## Nematology Newsletter Editors

Q. L. Holdeman
1956-1958

D. A. Slack
1959-1962

J. M. Ferris
1963-1965 (Co-Editor)

V. R. Ferris
1963-1965 (Co-Editor)

G. C. Smart
1966-1968 (Co-Editor)

R. P. Esser
1966-1968 (Co-Editor)

W. R. Jenkins
1969-1970

B. E. Hopper
1971-1976

E. M. Noffsinger
1977-1979

N. A. Lapp
1980-1982

G. R. Noel
1983-1986

E. C. Bernard
1987-1990

S. Spencer
1991-1993

P. Donald
1994-1997

E. C. McGawley
1998-1999

B. Sipes
2000-2002

## 104 Agriculture, Nematology and SON

P. Timper
2003-2005

B. Adams
2006-2008

# Appendix V
## Society of Nematologists Annual Meetings

| Year | Location | Description |
|---|---|---|
| 1962 | Corvallis, Oregon | Joint with AIBS* |
| 1963 | Amherst, Massachusetts | Joint with APS** |
| 1964 | Boulder, Colorado | First Independent Meeting |
| 1965 | Urbana, Illinois | |
| 1966 | Daytona Beach, Florida | |
| 1967 | Washington, DC | Joint with APS |
| 1968 | Columbus, Ohio | |
| 1969 | San Francisco, California | |
| 1970 | Washington, DC | |
| 1971 | Ottawa, Canada | |
| 1972 | Raleigh, North Carolina | |
| 1973 | Minneapolis, Minnesota | Joint with APS |
| 1974 | Riverside, California | |
| 1975 | Houston, Texas | Joint with APS |
| 1976 | Daytona Beach, Florida | |
| 1977 | East Lansing, Michigan | Joint with APS |
| 1978 | Hot Springs, Arkansas | |
| 1979 | Salt Lake City, Utah | |
| 1980 | New Orleans, Louisiana | |
| 1981 | Seattle, Washington | |
| 1982 | Knoxville, Tennessee | |
| 1983 | Ames, Iowa | Joint with APS |
| 1984 | Guelph, Canada | First International Nematology Congress |
| 1985 | Atlantic City, New Jersey | |
| 1986 | Orlando, Florida | Silver Jubilee of SON |
| 1987 | Honolulu, Hawaii | |
| 1988 | Raleigh, North Carolina | |
| 1989 | Davis, California | |
| 1990 | Veldhoven, The Netherlands | Second International Nematology Congress |

*AIBS - American Institute of Biological Sciences
**APS - American Phytopathological Society

| Year | Location | Description |
|---|---|---|
| 1991 | Baltimore, Maryland | |
| 1992 | Vancouver, Canada | |
| 1993 | Nashville, Tennessee | Joint with APS |
| 1994 | San Antonio, Texas | |
| 1995 | Little Rock, Arkansas | |
| 1996 | Goiser, Guadeloupe | Third International Nematology Congress Joint with IFNS[+] |
| 1997 | Tucson, Arizona | |
| 1998 | St. Louis, Missouri | |
| 1999 | Monterey, California | |
| 2000 | Quebec City, Canada | |
| 2001 | Salt Lake City, Utah | Joint with APS |
| 2002 | Tenerife, Canary Islands, Spain | Fourth International Nematology Congress Joint with IFNS |
| 2003 | Ithaca, New York | |
| 2004 | Estes Park, Colorado | |
| 2005 | Fort Lauderdale, Florida | |
| 2006 | Kauai, Hawaii | |
| 2007 | San Diego, California | Joint with APS |
| 2008 | Brisbane, Australia | Fifth International Nematology Congress Joint with IFNS |
| 2009 | Burlington, Vermont | Joint with Soil Ecology Society |

[+] IFNS - International Federation of Nematology Societies

# Appendix VI
# Society of Nematologists Constitution and Bylaws (2007 Version)

## Society of Nematologists CONSTITUTION
### ARTICLE I
### NAME

The name of this society shall be the Society of Nematologists, Inc. (Organized December, 1961)

### ARTICLE II
### PURPOSE

The purpose of the Society shall be the advancement of the science of nematology, in both its fundamental and its economic aspects. To serve this purpose, the Society shall act as an agency for the exchange of information, shall hold regular meetings, and shall promote and extend knowledge in all phases of the subject. The Society is organized and shall be operated on a nonprofit basis exclusively in advancing these educational and scientific purposes for the science of nematology.

## Bylaws

### ARTICLE I
### MEMBERSHIP

Section 1. CLASSES OF MEMBERSHIP. The classes of membership shall be Regular, Student, Emeritus, Honorary, and Sustaining Associate.

  a. REGULAR MEMBERS. Any person interested in the study of nematology is eligible for membership. A membership is activated upon receipt and acceptance of the applicant's dues.

  b. STUDENT MEMBERS. Any person who is studying nematology at the college or university level and who is enrolled in a recognized educational institution, may become a student member at reduced dues. Student membership may not exceed a period of five years and should be limited to individuals who are not gainfully employed more than two-thirds time. Applications must be endorsed by the major professor, and the employment status must be certified annually by the major professor or appropriate department head of the educational institution.

  c. EMERITUS MEMBERS. Upon approval by the Executive Board, regular members who are in good standing and who have retired from active service may, upon request, be continued as active members without payment of dues. These members shall be designated as emeritus members.

  d. HONORARY MEMBERS. Regular or emeritus members may be awarded an honorary life membership in recognition for meritorious and superlative contributions to the science of nematology.

   e. SUSTAINING ASSOCIATES. Any organization which contributes to the Society an amount prescribed in the Bylaws shall be designated a Sustaining Associate. That organization may designate one representative who shall have all the privileges of a regular member of the Society.

Section 2. PRIVILEGES. All individual members may participate in the annual meeting of the Society, vote on all matters submitted to the membership, hold an office (elective offices limited to regular members), appointment or committee membership, offer nominations for office, and publish in the Society's journals, when fulfilling the prescribed editorial reviews. Regular, Student and Honorary members as well as Sustaining Associates receive the journals and newsletter of the Society; Emeritus members retain all rights and privileges of regular membership, except being exempt from payment of dues, they will not receive any journal, unless they elect individually to do so at a rate 50 percent less than the regular membership rate. All members in good standing have the right to resign. Terminations of membership of any member for due cause is reserved by the Society, but excluding nonpayment of dues, no membership shall be terminated without an opportunity for a hearing before the Executive Board. The name of the Society shall not be used by any member or associate for financial gain.

Section 3. DUES AND FEES. Dues shall be paid annually, in advance, by January 1 of each year.

   a. The annual dues of a regular member shall be at a rate determined by the Executive Board each year.

   b. The dues of a student member shall be one-half of regular membership dues.

   c. Honorary and emeritus members shall be exempt from dues.

   d. The dues of a sustaining associate shall be at least $100 per year. Those who have not paid their dues by January 15, shall be notified that they are in arrears and that their names will be dropped from the roll, effective immediately. Members shall be reinstated upon payment of dues, plus a late fee, the amount to be determined by the Executive Board.

Section 4. SUBSCRIPTION AND BACK ISSUES. The price of nonmember subscriptions to the JOURNAL OF NEMATOLOGY, the price of single issues, and the price of back issues shall be determined by the Treasurer with approval from the Executive Board.

Section 5. ALLOCATION OF DUES AND FEES. A portion of the dues for regular and student members and contributions from sustaining associates, constitutes in full the annual subscription to the JOURNAL OF NEMATOLOGY and the NEMATOLOGY NEWSLETTER.

Section 6. MEETINGS OF MEMBERS.

    a. ANNUAL MEETING. There shall be an annual meeting of members to be held as designated by the Executive Board. Notice of the time and place of the annual meetings shall be given to all members not less than eight months prior to such meeting. The meeting site shall be selected to promote the attendance of as many members as possible.

    b. QUORUM. Those members present and entitled to vote at a meeting of the Society, after proper notice of the meeting, shall constitute a quorum.

    c. VOTING. At all meetings of the members, all motions shall be resolved by a majority vote of the members present in person except as specified elsewhere in the Bylaws.

    d. BUSINESS SESSIONS shall be conducted in accordance with Robert's Rules of Order, newly revised, in all cases to which the rules are applicable and in which they are not inconsistent with the Constitution and Bylaws and any special rules of order the Society may adopt. The President shall appoint a Parliamentarian to advise as needed; the Parliamentarian shall serve for the duration of the meeting for which the appointment is made.

    e. SPECIAL MEETINGS. Special meetings of members other than those regulated by statute, may be called at any time by a majority of the Executive Board. Notice of such a meeting, stating the purpose for which it is called, shall be served personally or by mail not less than sixty 60) days before the date set for such meeting. If the notice is mailed, it shall be directed to a member at the address as it appears on the membership roll. However, the giving of the notice stating the purpose of the meeting may be dispensed with for any meeting at which either all members shall be present or of which members not present have waived notice in writing. The Executive Board shall also, in like manner, call a special meeting of members whenever so requested in writing by members representing not less than one-third of the membership. No business other than that specified in the call for the meeting shall be transacted at any special meeting of the members.

### ARTICLE II
### OFFICERS

A.     SELECTION OF OFFICERS

Section 1. The officers of the Society shall be a President, a President-Elect, a Vice-President, a Secretary, and a Treasurer. All officers must be regular members of the Society.

Section 2. The Vice-President, Secretary, and Treasurer shall be elected by mail ballot, whereas the office of the President shall be filled by the elevation of the President-Elect from the preceding year. The office of President-Elect shall be filled by the elevation of the Vice-President from the preceding year. (In the event the President-Elect cannot succeed to the Presidency for any reason, an election for President shall be held at the same time and in the same manner as for the other elective offices. Similarly, should the Vice-President be unable to assume the office of President-Elect, an election for President-Elect shall be held as described for the other offices. The Secretary shall send a nomination ballot for those offices to be elected, to all members of the Society in time to allow the nominations to be returned not less than 120 days prior to the expiration of the term of office. The President shall appoint an Ad Hoc Elections Committee comprised of at least one Society member and one other person, to tabulate the ballots. The individual's name receiving the most nominations for each office will be on the final ballot for that office. The second nominee for each office shall be selected from among the five individuals who received the most nominations for the office by the membership after the individual receiving the most nominations. This selection will be made by a majority vote by the Executive Board. The final ballot shall be sent to all members not less than 60 days before expiration of the term of office. In circumstances where one person received sufficient ballots to qualify for more than one nomination, that member's name shall be placed on the final ballot as a candidate for the office for which the most nominations were received. In a circumstance where a person received an equal number of ballots that qualify that person for more than one office, that member's name shall be entered as a candidate for the office of the individual's preference. A plurality vote shall elect. No elected officer is eligible for immediate reelection to the same office.

Section 3. PRESIDENT, PRESIDENT-ELECT AND VICE-PRESIDENT (1-year terms). The Vice-President assumes office at the close of the Annual Business Meeting. The Vice-President shall serve for one year as Vice-President and the following year as President-Elect. The President-Elect assumes the office of President at the close of the Annual Business Meeting held at the end of the term of the incumbent President.

Section 4. SECRETARY. The Secretary shall serve for three years, beginning with the close of the Annual Business Meeting held at the end of the term of the incumbent Secretary.

Section 5. TREASURER. The Treasurer shall serve for three years, commencing with the close of the Annual Business Meeting held at the end of the term of the incumbent Treasurer. The terms of the Secretary and Treasurer terminate in alternate years.

Section 6. APPOINTED EXECUTIVE BOARD MEMBERS. Each of the three appointed Directors of the Executive Board serves a 2-year term. Each President-Elect, prior to being installed as President at a given annual meeting, shall have the privilege and obligation to fill the vacancies created by appointed Directors whose terms will expire during the annual meeting. The incoming appointed Director will assume office at the close of the Annual Business Meeting for that year. The Editor-in-Chief of the JOURNAL OF NEMATOLOGY, Editor of the NEMATOLOGY NEWSLETTER, Web Site Editor, and Public Relations Representative upon being installed in these appointive offices, shall become Directors of the Executive Board.

B. DUTIES OF OFFICERS

Section 1. PRESIDENT, PRESIDENT-ELECT, AND VICE-PRESIDENT. The President shall preside over meetings of the Society and the Executive Board, and perform such other duties as may be necessary. The President shall preside at the annual meeting of the Society and of the Executive Board and report on the condition of the business and affairs of the Society. All regular and special meetings of the members and the Executive Board shall be called by the President, in accordance with these Bylaws and the Constitution. The President shall appoint and remove, employ and discharge, and fix the compensation of all assistants, agents, employees and clerks of the Society other than the fully appointed officers subject to approval of the Executive Board; see that the books, reports, statements, and certificates required by the Statutes are properly kept, made, and filed according to law; serve as an ex-officio member of all Committees of the Society. The President shall enforce all of the duties which are required by the law and incident to the position and office.

The President-Elect shall preside in the absence of the President and perform the functions of the President when so acting. The President-Elect serves as chairman of the Program Committee for the next annual meeting of the Society. In this capacity, the President-Elect shall preside at all meetings of the Program Committee; report to the Executive Board on all program plans that require their approval; coordinate the development of the Annual Meeting Program, including the editing of abstracts in cooperation with the Editor-in-Chief of the JOURNAL OF NEMATOLOGY; appoint all Chairs and Co-Chairs for contributed paper sessions; and coordinate with the Chair of the Local Arrangements Committee all arrangements for the program. The President-Elect shall appoint the new members of standing committees at least 30 days prior to the annual meeting at which he/she is installed as President in order that new members can participate in committee meetings during the annual meeting.

The Vice-President is a member of the Executive Board and presides at all Board meetings in the absence of both the President and President-

Elect. The Vice-President shall serve as the Chair of the Meeting Site Selection Committee, and as Vice-Chair of the Program Committee. Immediately after taking office, the Vice-President commences planning the program for the Annual Meeting at which he/she shall be installed as President.

Section 2. SECRETARY. The Secretary shall take and maintain the minutes of the annual meetings of the Executive Board and of the membership; prepare and mail the call for nominations, election Ballots, constitutional amendment proposals, and other notices of the Society; maintain annual reports of officers and committees, and forward such records to the Archivist as appropriate at the end of each year. The Secretary will bring to have accessible for presentation to the Executive Board at their stated meetings, all communications addressed to the Secretary officially by the President or any officer or member of the Society; attend to all correspondence and perform all the duties incident to the office of Secretary; and announce results of the election of officers in appropriate news media; and assist the President in preparing the agenda for regular and called meetings of the Executive Board and of the membership. The Secretary shall maintain and update the official file copies of the CONSTITUTION & BYLAWS and the MANUAL OF OPERATIONS of the Executive Board, Committee Chairs, and Representatives.

Section 3. TREASURER. The Treasurer shall have care and custody and be responsible or all the funds and securities of the Society and deposit all funds in the name of the Society. The Treasurer shall sign, make, and endorse in the name of the Society, all checks, drafts, warrants, and orders for payment of money, and pay out and dispose of same and receipt thereof, as authorized by the President or the Executive Board; exhibit at all reasonable times, the Treasurer's books and accounts to any Board Member or Member of the Society upon application at the office of the Treasurer; render a statement of the condition of the finances of the Society at each meeting of the Executive Board and at such other times as shall be required, and a full financial report to the membership at the Annual Meeting of the Society. The Treasurer and/or the Business Office of the Society shall keep correct books of accounts of all business transactions, and such other books of account as the Executive Board may require; keep the membership roll so as to show at all times the names of members, alphabetically arranged, their respective places of residence, their post office addresses, and keep such records open, subject to the inspection of any member of the Society and permit such members to make abstracts from said records to the extent prescribed by law. The Treasurer will provide access to the financial records for an audit by a Certified Public Accountant and perform all duties pertaining to the office of Treasurer. In consultation with the Executive Board, the Treasurer shall prepare an annual budget for the Society, including all costs of all publications, for consideration and approval of the Executive Board during the annual meeting of the Society; and make recommendations to the Executive Board on all matters pertaining to the immediate and long-term financial planning for an efficient management of the Society affairs. The Treasurer shall assist the Program Chair (President-Elect) and the Chair of the Local Arrangements Committee in developing a budget for the Annual Meeting of the Society which is approved by the Executive Board.

## ARTICLE III
## THE EXECUTIVE BOARD/BOARD OF DIRECTORS

Section 1. DUTIES OF EXECUTIVE BOARD. The Executive Board, comprised of thirteen Directors, shall have the control and general management of the affairs and business of the Society; this Board may adopt such rules and regulations for the conduct of its meetings and management of the Society as it deems proper, subject to these Bylaws and the laws of the State of Maryland.

Section 2. COMPOSITION. The Executive Board shall include the President, President-Elect, Vice-President, Secretary, and Treasurer, who are designated as the Board of Directors, the three appointed members of the Society, the immediate Past-President, the Editor-in-Chief of the JOURNAL OF NEMATOLOGY, and the Editor of the NEMATOLOGY NEWSLETTER.

Section 3. REGULAR EXECUTIVE BOARD MEETINGS. Regular meetings of the Executive Board shall be held at the time of the Annual Meeting of the Society. The President presides at all Board meetings; or in the President's absence, the President-Elect presides. In the absence of both the President and the President-Elect, the Vice-President presides at meetings of the Executive Board.

Section 4. SPECIAL EXECUTIVE BOARD MEETINGS. Special meetings of the Executive Board may be called by the President at any time, and shall be called by the President or the Secretary upon the written request of the majority of the Board. Notice of meetings, other than the regular annual meetings, shall be given to all Board Directors in person, or by mailing to their last known post office addresses at least 30 days before the date therein designated for such meeting, including that day of mailing, a written or printed notice thereof specifying the time and place of such meeting, and the business to be brought before the meeting; no business other than that specified in such notice shall be transacted at any special meeting. Any business of the Society may be transacted by the Executive Board without prior notice if every Board Director is present.

Section 5. QUORUM. At any meeting of the Executive Board, the majority of the Directors (7) shall constitute a quorum for the transaction of business.

Section 6. VOTING. Each of the thirteen Directors of the Executive Board shall have one vote.

Section 7. VACANCIES. If the President is unable to continue serving the Society, the office shall be filled for the remainder of the term by the President-Elect. Other vacancies on the Executive Board that occur between Annual Meetings shall be filled by appointment for the unexpired term by a majority of the remaining Board Directors. The Executive Board shall remove, by a two-thirds vote, any Director of the Board for failure to carry out the responsibilities of office.

Section 8. POLICY. Between Annual Meetings of the Society, the Executive Board must carry out the necessary functions to implement existing policies of the Society. The Executive Board is authorized to conduct mail ballot(s) of the Society membership between annual meetings of the Society on matters the Board deems necessary to place before the entire membership for a vote, provided each ballot is accompanied with explanatory information. Interim actions on policy are reported to the membership at the next Annual Meeting of the Society.

### ARTICLE IV
### COMMITTEES AND SOCIETY REPRESENTATIVES

A. GENERAL

Society Committees are of three categories: Standing Committees, Special Committees, and Ad Hoc Committees. Unless indicated differently in the Bylaws, the President-Elect, as incoming President, designates the Vice-Chair, if any, of each committee which person will elevate to Chair of the Committee in the following year, unless circumstances require an alternate process which then must be approved by the Executive Board.

B. STANDING COMMITTEES

Section 1. The Standing Committees to be designated in the MANUAL OF OPERATIONS are concerned with matters of continuing interest to the Society, including general policies and program development. Each committee must submit to the Executive Board an annual report which is published, at least in abstract form, as part of the official report of the Annual Meeting of the Society. Upon recommendation by the Executive Board, Standing Committees may be added or discontinued by a plurality vote of the membership at the Annual Meeting.

Section 2. TERMS OF OFFICE AND ROTATION. Unless approved otherwise by the Executive Board, each member of a Standing Committee shall serve for a period of three years, commencing with the close of the Annual Meeting of the Society. Appointments shall be so arranged that approximately one-third of the terms shall expire annually.

Section 3. APPOINTMENT OF STANDING COMMITTEES AND FILLING OF VACANCIES.

a. The President-Elect, as incoming President, shall appoint the new members of all Standing Committees at least 30 days prior to the Annual Meeting.

b. The President shall fill vacancies in Standing Committees by appointment. The appointees shall serve until the terms expire or successors are appointed to serve for the remainder of the unexpired terms, whichever comes first.

Section 4. COMPOSITION OF COMMITTEES. The Program Committee shall consist of the Chair (President-Elect), Vice-Chair (Vice-President), and the Chairs of the following committees: Biological Control of Nematodes, Ecology, Education, Extension, Industry, Plant Resistance to Nematodes, Regulatory, and Systematic Resources. The Vice-President shall serve as Chair of the Meeting Site Selection Committee. The immediate Past-President shall serve as Chair of the Long-Range Planning Committee with responsibilities of consulting and communicating with the Chairs of appropriate subject-matter committees in this regard. The President, President-Elect and/or Executive Board shall determine the composition of all other committees.

C. SPECIAL COMMITTEES
The special committees are concerned with matters of short-term interest to the Society. A Special Committee will be appointed for a period which in the judgment of the Executive Board is required to fulfill its purpose. Progress reports should be provided and a final report must be made to the Executive Board at the termination of the Committee's appointment.

D. AD HOC COMMITTEES
These committees are concerned with issues of specific purpose involving a limited period. The President of the Society establishes these committees and appoints their membership and Chairs. The term of any Ad Hoc Committee will not extend beyond the term of the President who appoints it. The succeeding President, however, may reinstate an Ad Hoc Committee. These committees make reports to the President or the Executive Board as requested by the President.

E. SOCIETY REPRESENTATIVES
The President-Elect, as incoming President, shall appoint all Society Representatives to other organizations, such as, CAST, AIBS, ASC, etc., prior to the Annual Meeting of the Society.

**ARTICLE V**
**PUBLICATIONS**

Section 1. THE SOCIETY SHALL ISSUE JOURNALS AND NEWSLETTERS,

AND MAINTAIN A WEB SITE, as determined by the Executive Board of the Society with the approval of a plurality of the voting membership. The types of publications and frequency of issues shall be determined by the Executive Board.

Section 2. JOURNALS. For each Journal, the Executive Board shall appoint an Editor-in-Chief. The Editor-in-Chief, subject to approval by plurality vote of the Executive Board, shall appoint Editorial Board Members, and as many Editors and Associate Editors as deemed necessary. A Senior Editor will be appointed by the Executive Board one year prior to the completion of the Editor-in-Chief's term of office. The Editor-in-Chief shall serve for a term of three years; and may be reappointed thereafter, for two consecutive 1-year terms at the discretion of the Executive Board. The Editorial Board Members shall each serve a term of three years, terms not to expire concurrently, and may be reappointed. Associate Editors shall serve terms of two years with the terms of one-half expiring in even years and the other half expiring in odd years, and may be reappointed.

Associate Editors for the ANNALS OF APPLIED NEMATOLOGY shall serve three-year terms. If the office of any editor is vacated for any reason, the President and the Editor-in-Chief shall jointly nominate a replacement. Where possible, the newly named Editor-in-Chief should have served one year as Senior Editor prior to assuming the position of Editor-in-Chief. The duties of the Editor-in-Chief, Editorial Board Members, Editors, and Associate Editors, shall be those necessary to maintain a high quality journal. Each Editorial Board Member shall have an equal voice and equal vote in all Journal matters not reserved for other Society Officers by the Constitution and Bylaws.

Section 3. NEWSLETTER. The Executive Board shall appoint an Editor for the Newsletter. The Editor may request a Co-Editor or Associate Editors. The Editor of the Newsletter shall serve for a term of three years, and may be reappointed thereafter, for two consecutive 1-year terms at the discretion of the Executive Board. Any Co-Editor or Associate Editor shall serve terms no longer than that of the Editor. The duties of the Editor shall be those necessary to keep Society members aware of current events in and affecting the science of nematology and related disciplines.

Section 4. SPECIAL PUBLICATIONS. The Society may solicit and/or receive manuscripts for consideration as special publications. Acceptance for special publication shall be the responsibility of the Executive Board, subject to the following conditions: (a) a Special Committee shall be appointed by the Executive Board to study the feasibility of publication of each manuscript and their findings shall be reported to the Executive Board; (b) if the manuscript is accepted for publication, the Special Publications Committee shall handle all pre-publication and post-publication publicity related to sales, and shall serve for one year after the publication is released; and (c) actual sales of such publications shall be the responsibility of the Treasurer or Business Manager.

If the Executive Board accepts a manuscript for special publication, a Special Editor shall be appointed for that publication. The Special Editor, in association with the authors, shall decide on format, type of printing, paper, binding, cover, number of copies, cost, and other publishing details. At least two bids shall be obtained. The Special Editor shall report to the Special Publication Committee and the Executive Board. All costs relating to each special publication must be approved by the Executive Board and will be borne by the Society; however, special publications shall be managed so as to recover the cost through contributions and/or sales of individual copies.

Section 5. The Executive Board shall appoint a Web Site Editor to maintain the official Society Web Site. Content of the Web Site shall be determined by the Executive Board. The Web Site Editor shall serve for a term of three years.

Section 6. The Society shall issue such other publications, brochures, and publicity materials as deemed necessary by the Executive Board. The Executive Board shall decide the manner of funding of such issues.

### ARTICLE VI
### HONORS

Section 1. HONORARY MEMBER. This award, the highest honor the Society bestows, is limited to no more than one person per year. The number of living persons in this category of membership shall not exceed 2% of the Regular and Emeritus membership counted at the time of the Annual Meeting prior to the time of selection. Nominations for Honorary Member(s) may be made by any member of the Society. The nominator must provide a dossier for each individual nominated to the Honors and Awards Committee at least 180 days prior to the Annual Meeting of the Society. The Honors and Awards Committee will evaluate all nominations for Honorary Member. This Committee will forward their nomination and dossier to the Executive Board no later than 120 days prior to the Annual Meeting. The Executive Board must give approval of the nominee for Honorary Member of the Society at least 90 days prior to the Annual Meeting. The President will notify the Honorary Member-Elect, at least 60 days prior to the Annual Meeting.

Section 2. FELLOW. This honor is bestowed upon regular and emeritus members in recognition of distinguished contributions to the science of nematology. No more than 0.40% of the membership may be elected to Fellow of the Society in any given year. Nominations for Fellow may be made by any member of the Society. All nominations for Fellow must include a dossier which must reach the Honors and Awards Committee at least 180 days prior to the Annual Meeting of the Society. The selection of Fellows is made by the Honors and Awards Committee. The President will notify the Fellows-Elect no later than 60 days prior to the Annual Meeting. The Executive Board shall be advised of the Fellows-Elect before the public announcement at the Annual Meeting.

Section 3. SPECIAL HONORS. The Executive Board may establish other special honors and awards categories it deems appropriate.

## ARTICLE VII
## FUNDS

Section 1. CONTROL. The control of Society funds from dues, subscriptions, publications, gifts, endowments, bequests, and investments is vested in the Executive Board to be managed through the Treasurer and any appointed Executive Officer(s) and/or committees. The Executive Board shall have the authority, in its discretion, to take such actions, if any, as they deem necessary or appropriate for the safekeeping of the Society's cash, cash accounts, and investments, including, but not limited to, the bonding of signatories on Society accounts or requiring multiple signatures for withdrawals.

Section 2. OBLIGATIONS. The Executive Board is authorized to enter into obligations and to pay obligations essential to conducting the affairs of the Society and the editing, printing, and publishing of the official journals and other publications authorized by the Society.

Section 3. ANNUAL AUDIT. Receipts and disbursements of the Society shall be examined annually by a Certified Public Accountant (CPA) to be approved by the Executive Board. The CPA shall provide an audit of all accounts as requested by the Executive Board. The reports of the Treasurer and/or Business Manager, and the Certified Public Accountant shall be published annually.

## ARTICLE VIII
## AFFILIATED SOCIETIES AND ORGANIZATIONS

The Executive Board is authorized to investigate and recommend to the Society, the establishment of affiliations with other societies and organizations. Such affiliations should be developed for the purpose of promoting national and international collaboration between the Society of Nematologists, other societies of nematologists, and other appropriate organizations.

## ARTICLE IX
## RATIFICATION AND AMENDMENTS

Section 1. This Constitution and Bylaws become effective upon their ratification according to the procedure for amending given in Article VII of the August 1980 revision of the Constitution and Bylaws of the Society. This

1993 revision supersedes the Constitution and Bylaws as originally ratified in February 1962 and all intervening revisions.

Section 2. Proposals for amendments to this Constitution and Bylaws may be made to the Executive Board by any member of the Society.

Section 3. Changes in SECTION 3 THROUGH 5 OF ARTICLE I OF THE BYLAWS may be made by a two-thirds majority vote of the members in attendance at the Annual Meeting of the Society, provided that the proposed amendment(s) were distributed to the membership at least 60 days prior to the date of the Annual Meeting at which such proposals are to be considered.

Section 4. THIS CONSTITUTION AND THE REMAINING SECTIONS OF ARTICLE I AND ALL OTHER ARTICLES OF THE BYLAWS may be amended under the following provisions:

   a. Any proposals for amendments shall first be submitted in writing to the Secretary of the Society at least 90 days before the Annual Meeting of the Society at which action thereon is to be taken.

   b. The Executive Board shall consider all amendment proposals which may be modified by the Executive Board only to clarify obscure language or to facilitate carrying out the intent of the proposed amendment(s).

   c. At least 60 days before the annual meeting at which action is to be taken, the Executive Board shall mail to all members a copy of the proposed amendment(s). This notification may be by publication in the NEMATOLOGY NEWSLETTER, or by letter to each member of the Society.

   d. Proposals to amend the Constitution, duly made in the prescribed manner, together with the recommendations of the Executive Board, shall be presented to the members of the Society at the Annual Business Meeting for discussion and vote as described in paragraphs e-g of this section.

   e. Approval of a proposed amendment requires an advisory approval by a two-thirds majority of members voting in a mail ballot. If a proposed amendment receives advisory approval at the Annual Meeting, it shall be published in the next issue of the NEMATOLOGY NEWSLETTER with the part of the Constitution affected by the proposed amendment.
   Justification and criticism for the proposed amendment(s) and the recommendations of the Executive Board concerning its passage shall be printed with the proposal. Should the Executive Board be opposed to the proposed amendment, supporting comments of the group originating the amendment shall be included. A mail ballot shall be sent to the entire membership within 30 days after the proposed amendment is published in the NEMATOLOGY NEWSLETTER.

   f. The President shall appoint a special committee of two members (or the

Secretary and one other person) to tabulate the ballots with the Secretary serving as Chair of this Committee. The Committee shall tabulate all votes received 60 days after distribution and inform the President in writing and the members through the NEMATOLOGY NEWSLETTER of the results of the ballot. The amendments shall be adopted if two-thirds or more of the votes are in favor, otherwise, it shall be defeated.

g. An adopted amendment shall become effective as of the date the ballots are counted, unless a later date is specified in the amendment itself.

## ARTICLE X
## DISSOLUTION

In the event that the Society is dissolved for any reason, all assets remaining after payment of debts shall be distributed exclusively for educational and scientific purposes as recommended by the Executive Board.

## APPENDIX
## Articles of Incorporation
## (Without authorized capital stock)
## SOCIETY OF NEMATOLOGISTS, INC.

FIRST; We, the undersigned, Charles A. Dukes, Jr., whose post office address is 5303 Baltimore Avenue, Hyattsville, Maryland 20781, Jo B. Fogel, whose post office address is 5303 Baltimore Avenue, Hyattsville, Maryland 20781, and Joan Katz, whose post office address is 5303 Baltimore Avenue, Hyattsville, Maryland 20781, each being at least twenty-one years of age, do hereby associate ourselves as incorporators with the intention of forming a corporation under and by virtue of the General Laws of the State of Maryland.

SECOND; The name of the corporation (which is hereinafter called the Corporation) is Society of Nematologists, Inc.

THIRD; The purposes for which the Corporation is formed are as follows: a) To advance the science of nematology, in both its fundamental and its economic aspects; to act as an agency for the exchange of information; to hold regular meetings and promote and extend knowledge in all phases of nematology, b) and generally to carry on any other business in connection therewith not contrary to the laws of the State of Maryland, and with all the powers conferred upon non-profit corporations which are contained in the General Laws of the State of Maryland.

FOURTH; The Post Office address of the principal office of the Corporation in this State is 5303 Baltimore Avenue, Hyattsville, Maryland. The name and post office address of the resident agent of the Corporation in this State are Charles A. Dukes, Jr., 5303 Baltimore Avenue, Hyattsville, Maryland. Said resident agent is a citizen of this State and actually resides herein.

FIFTH; The Corporation is not authorized to issue capital stock.

SIXTH; The number of directors of the Corporation shall be not more than nine which number may be increased or decreased pursuant to the Bylaws of the Corporation, but shall never be less than three; and the names of the directors who shall act until the first annual meeting or until their successors are duly chosen and qualified are Harold A. Jensen, Burton Y. Endo, and George A. Bird.

SEVENTH; The duration of the Corporation shall be perpetual.

# Appendix VII
# Nathan A. Cobb Nematology Foundation Bylaws (Original Version)

## BYLAWS
## OF THE NATHAN A. COBB NEMATOLOGY FOUNDATION, INC.
### A NON-PROFIT CORPORATION
### INCORPORATED IN THE STATE OF KANSAS, USA

### ARTICLE ONE - ORGANIZATION

The name of this Organization shall be: The Nathan A. Cobb Nematology Foundation, a non-profit Florida U.S.A. Corporation.

The Foundation shall have a seal which shall be in the following form:

The Foundation may at its pleasure by a majority vote of the membership change its name at the annual meeting provided written notice of such change is mailed to members of record at least 60 days prior to the meetings.

### ARTICLE TWO - PURPOSES

The purpose of this Foundation would be similar to those of scientific foundations and would include but is not limited to the following:

Provide grants to students for study, for travel and to participate in annual meetings of nematological and related societies.

Provide assistance to scientists to encourage creative thought and to assist in the early development of innovations in Nematology and their applications.

Provide grants to plan, develop, and conduct special workshops, courses and programs on topics of interest to nematological scientists and students throughout the world.

Provide grants to students and scientists to help defray costs of publications of nematological interest.

Provide assistance for development and publication of books and other publications of Nematological interest.

### ARTICLE THREE - MEMBERSHIP

Membership in this Foundation shall be automatic to all who are Regular, Student, Emeritus or honorary members of the Society of Nematologists, Inc., a Maryland Corporation, and/or designated representatives of entities making contributions to the Foundation as prescribed from time to time by the Board of Directors.

All members shall be entitled to vote on issues and other wise have equal rights in the function of the Foundation.

### ARTICLE FOUR - MEETING

The annual membership meeting of this Foundation shall be held in conjunction with the annual, general business meeting of the Society of Nematologists, a non-profit Maryland Corporation. Those members present at the annual meeting shall constitute a quorum to transact business of the Foundation. At the meeting, members shall have the opportunity to discuss and vote on matters pertinent to affairs of the Foundation.

Special meetings of the Foundation or the Board of Directors may be called by the Chair when it is deemed in the best interest of the Foundation. Notices of such meeting shall be mailed to members at their addresses as they appear in the membership roll book at least 60 days before the scheduled date set for such special meeting. Such notice shall state the reasons that such meeting has been called and the business to be transacted at such meeting and by who called. At the request of 3 members of the Board of Directors or 10 regular members, the Chair shall cause a special meeting to be called but such request must be made in writing to the Chair at least 75 days before the requested date. No other business but that specified in the notice may be transacted at such special meetings.

### ARTICLE FIVE - VOTING

Votes taken by the Board of Directors and the membership may be by voice or secret ballot. Secret ballots will be taken if upon voice vote 20 percent of members present request such vote. At any annual, regular or special meeting if a quorum is present as outlined elsewhere, any question may be voted upon in the manner and style provided above.

At all votes by secret ballot, the Chair of such meeting shall appoint a committee of three who shall act as Inspectors of Election and who shall at the conclusion of such balloting certify to the Chair the results and provide a written certified copy which shall be physically affixed in the mount book to the minutes of that meeting.

No Inspector of Election shall be a candidate for office or shall be personally interested in the question voted upon.

### ARTICLE SIX - GENERAL ORDER OF BUSINESS

Reading of the minutes of the preceding meeting; Report of officers; Reports of committees; Old and unfinished business; New business; General good and welfare.

### ARTICLE SEVEN - BOARD OF DIRECTORS

During its first year of existence, the business of this Foundation shall be managed by the Executive Board of the Society of Nematologists, Inc., a Maryland Corporation. Thereafter, the business of this Foundation shall be managed by a Board of Directors consisting of eight members. Four Directors shall be directly elected by the membership and the remaining four Directors shall be officers (President, President-elect, Secretary and Treasurer) of the Society of Nematologists, Inc., a Maryland Corporation.

The Secretary of the Society of Nematologists, Inc., a Maryland Corporation, shall conduct the election of those Directors who are not also members of the Executive Board of that Society. These elections shall be conducted concomitantly with elections of officers of the Society of Nematologists, Inc., a Maryland Corporation and shall follow the procedures described in Section 2, Article II-A of the bylaws of that Society.

The Board of Directors shall have full control and management of the affairs, funds and business of the Foundation. The Directors may expend interest moneys only, while expenditures of principal moneys can only be approved by a majority of the membership at the annual meeting. Such Board of Directors shall only act in the name of the Foundation when it shall be regularly convened by its Chair after due notice to all the Directors of such meeting.

Meetings of the Board of Directors shall be held during the annual meeting of the Society of Nematologists, Inc., a non- profit Maryland Corporation. Those Directors preset shall constitute a quorum. Each Director shall have one vote and voting may not be done by proxy. Voting on specific issues, however, may be conducted by mail ballot as appropriate.

The Board of Directors may make such rules and regulations covering its meetings as it may in its discretion determine necessary.

Vacancies in the Board of Directors shall be filled by a vote of the majority of the remaining members of the Board of Directors for the balance of the remaining term.

A Director may be removed by two-thirds majority of the Board of Directors when sufficient cause exists for such removal. The Board of Directors may entertain charges against any Director. A Director shall have the right to be represented by counsel upon any removal hearing. Upon a simple majority vote, the Board of Directors shall adopt such rules as it may in its discretion consider necessary for the best interests of the Foundation for this hearing.

## ARTICLE EIGHT - OFFICERS

Beginning with its second year of existence, the officers of the Foundation shall be as follows: Chair, Vice-Chair, Secretary and Treasurer. The Chair of the Foundation by virtue of the office shall be Chair of the Board of Directors. Similarly, the Vice-Chair, Secretary and Treasurer of the Foundation by virtue of their office shall be the same officers for the Board of Directors. Officers shall be directly elected by the membership.

Members of the Executive Committee of The Society of Nematologists, Inc., a Maryland Corporation, shall be ineligible to serve as officers of the SON Foundation.

The Chair and Vice-Chair of the first elected Board of Directors shall serve for a term of 4 years and the Treasurer and Secretary shall serve for a term of 3 years. Thereafter, all of these 4 directors shall be elected to serve for a term of 3 years.

The Chair shall preside at all membership and Director meetings and present at each annual meeting of the Foundation and annual report of the work of the Foundation, appoint all committees, temporary or permanent, see that all books, reports and certificates as required by law are properly kept or filed, be one of the officers who may sign checks or drafts of the Foundation, and have such powers as may be reasonably construed as belonging to the Chief Executive of any organization.

The Vice - Chair shall in the event of the absence or inability of the Chair to exercise his or her office become Acting Chair of the Foundation with all the rights, privileges and powers as if he or she had been the duly elected Chair for the remaining term of the Chair, as applicable.

The Secretary shall keep the minutes and records of the Foundation in appropriate books, file any certificate required by an statute, federal or state, give and serve all notices to members of the Foundation, be the official custodian of the records and seal of the Foundation, present to the membership at any meetings any communication addressed to the Secretary of the Foundation, submit to the Broad of Directors any communications which shall be addressed to the Secretary of the Foundation, attend to all correspondence of the Foundation and exercise all duties incident to the office of Secretary.

The Treasurer shall have the care and custody of all moneys belonging to the Foundation, be solely responsible for such moneys or securities of the Foundation and be one of the officers who shall sigh checks or drafts of the Foundation. No special fund may be set aside that shall make it unnecessary for the Treasurer to sign the checks issued upon it. The Treasurer shall render at the annual meeting and at other stated periods as the board of Directors shall determine a written account of the finances of the foundation and such report shall be physically affixed to the minutes of the Board of Directors of such meeting and shall exercise all duties incident to the office of Treasurer.

No officer or Director shall for reason of the office be entitled to receive any compensation other that expenses, but nothing herein shall be construed to prevent an Officer or Director from receiving any compensation from the Foundation for duties other than as a Director or officer.

All officers-elect shall assume the duties of the office at the end of the annual meeting of the Foundation.

### ARTICLE NINE - SALARIES

The Board of Directors shall hire and fix the compensation of any and all employees which they in their discretion may determine to be necessary in the conduct of the business of the Foundation.

### ARTICLE TEN - COMMITTEES

The permanent committees shall be: Donor Relations and Special Projects. The Board of Directors shall appoint additional permanent and ad hoc committees as they deem appropriate.

### ARTICLE ELEVEN - AMENDMENTS

These Bylaws may be altered, amended, repealed or added to by an affirmative vote of not less than 51% of the regular members present at the annual meeting provided written notice is mailed to such members no less that 60 days prior to the meeting.

### ARTICLE TWELVE - DISSOLUTION

In the event that this Foundation is dissolved for any reason, all assets remaining after payment of debts shall be distributed for exclusively educational and scientific purposes as recommended by the Board of Directors.

# Appendix VIII
## Nematological Quotes

Numerous interviews were conducted while preparing to write this volume. Each interview contained interesting and insightful information. Unfortunately, we could not include all of the "words of nematological wisdom" into the pages of this book. We would therefore like to present some of these quotes which did not make the final cut.

---

**K. R. Barker**, SON President (1979-1980): "At North Carolina State University, for example, there are a lot of students working on nematodes. Some of them may be majoring in genetics or microbiology or a combination of them while others may be majoring in plant pathology, genetics or biochemistry. Those working in the new facets of ecology may need to co-major in plant pathology or nematology and some aspects of ecology so that using Van Gundy's 1980 challenge of 'nematologists must take their blinders off' in pursuit of fundamental problems. Nematology has moved into a new era but rather than looking at our own narrow interests, as was done in the 1950's-1970's, these new scientists coming into the field must be able to pursue broader research goals and visions."

**K. R. Barker**, SON President (1979-1980): "The Society of Nematologists was the embodiment of the vision of J. N. Sasser, M. Allen, D. Raski, S. Sher and D. P. Taylor to have this comprehensive discipline of nematologists that included nematodes of all aspects which were traditionally addressed as animal parasites by parasitologists, to some degree the veterinary science aspects of nematology, to come together to form a major discipline."

**H. Ferris**, nematologist, University of California-Davis: "I remember the meeting in Minneapolis, St. Paul in 1972. It might have been the first meeting I was involved with and attended. I was in awe of Gerald Thorne and people I felt obliged to address as 'Sir' because they were such huge figures in terms of my academic background. I have attended regularly since then. What issues being faced, I do not know. It seemed like nematicides had been a big driving factor. I can remember being a little dismayed that a lot of the talks at the meetings where on nematode control. There was the guy from California who talked about the 5 gallons per acre. The guy from Colorado would talk about the 5-6 gallons per acre. I was thinking, 'is this what it is all about?' DBCP at 5-6 gallons per acre? Is there anything else to it?"

**H. Ferris**, nematologist, University of California-Davis: "Concerning technicians, they can affect the institutional memory, the progress made in the department and the training of graduate students. One of the best things that ever happened to me in my career was that I worked in the same lab as DeWitt Byrd, Jr. He was Dr. Nusbaum's technician for many years. Dr. Nusbaum was working with Ken Barker to develop a pilot diagnostic service. It was called the Pilot Program for North Carolina. At that time, I was a graduate student hanging around the lab. I was able to participate in the development of techniques. At the same time, Dr. Nusbaum worked with Furnie Todd, who was working on his 'Research on Wheels' with the tobacco plant pathology program and nematology program across the state of North Carolina. Todd's 'Research on Wheels' (consisted of farm tours) during the growing season followed by a symposium with all the participating growers. It was a wonderful system. I do not know of any extension program in the world that was the equivalent to what that was. One of my responsibilities as a research assistant was to participate with DeWitt and other technicians in processing and identifying the samples that came in from Todd's 'Research on Wheels' plots. We had to process and identify the samples before July 11 because the book had to be ready for the symposium. I remember a couple of times when we stayed there all night counting nematodes. I remember falling asleep on the microscope. The enthusiasm came from people like Dewitt. He lit the fire in my belly for that kind of stuff."

**I. King**, graduate student at the University of California-Riverside: "My undergraduate program [at the Memorial University of New Finland in Canada] was focused on organismal biology. I was there as part of the committee that changed the undergraduate program from a theoretical concept to genetics and evolution. I think about nematology in that sense. I do not know much about the plant-parasitic side of things."

**S. Lewis**, SON President (1990-1991): "There were not as many hires based on a particular nematode problem. The hires across the state of South Carolina were related to discipline oriented programs that needed a nematologist as part of a research program."

**T. Niblack**, SON President (2005-2006): "Nematologists are rarely trained as nematologists. They are trained as some kind of specialists usually at the molecular level. All have to have a molecular component to get through school anymore."

**C. H. Opperman**, molecular nematologist, North Carolina State University: "The 1980's saw an explosion of research and discovery on basic biological mechanisms. Genetic engineering came of age; it was possible to manipulate the most basic components of a living organism to improve its performance in the field without application of costly and potentially hazardous chemicals."

**C. H. Opperman**, molecular nematologist, North Carolina State University: "It is very hard to see the future because we haven't really made it yet. I do not know what technology is around the corner. There is lots of stuff going on that is good. I think five years from now, its going to be back to the heavy days of having lots of products in the developmental stages."

**J. N. Sasser**, SON President (1963-1964): "I think what made the Society of Nematologists be established and grow was that there were a number of dedicated people doing their best to make a contribution to their science. I was amazed how fast the Society grew."

**D. P. Schmitt**, SON President (1989-1990): "The methods Cobb developed in the early 1900's are not drastically different from the methods used today. Nematologists are still using sieves and funnels. An elutriator is basically just sieves and funnels hooked up to a $20,000 machine. The only difference is some individual is not mixing by hand."

**J. L. Starr**, SON President (1996-1997): "When I came to Texas A&M, and the ones that I graduated with were almost the last ones to experience this, the departments had departmental assistantships for every student; every faculty member also had a technician and a level of support."

**A. C. Tarjan**, Committee of Nine member: "The Society will continue to exist although people will be leaving it as there are splinter groups breaking off going in other directions because 'birds of a feather flock together.' I always feel that basic nematology will be the glue to hold us together. I am a nematologist. I should belong to the Society of Nematologists. Nematodes are the main subject. If you are working with nematodes, you ought to be in the Society."

**S. Van Gundy**, SON President (1973-1974): "People live in an urban society and go down to the market and buy food. It is there. It is cheap. They do not worry if you have a nematode problem or a *Verticillium* problem or a *Fusarium* problem. They are more interested in doing social things. I think we are in an era were we are dealing with socialization and social problems, not with agricultural problems."

**J. M. Webster**, SON President (1982-1983): "The Society of Nematologists will not disappear as long as there are nematologists around who want to talk. There is a certain common thread between nematologists. They like to talk about the aspects of their discipline. It does not matter if they are talking about the wrinkles on the cuticle on the structure or weather they are talking about molecular aspects. The excitement also is linking up with others. Nematologists are a group of people working with nematodes."

**J. M. Webster**, SON President (1982-1983): "The value of our science is important in contemporary agriculture but its image is not as strong as it should be. There is a tendency for nematology to have a 'soft' image associated with a 'muck and shovel' program. The truth is, however, that like the virology, medicine and the pharmaceutical industries, we have the opportunity in nematology to take full advantage of some of the new advances in biotechnology. Even in nematology, we must not discount the essential value of some of the more basic science approaches and achievements in helping us resolve some of the more applied problems."

*Index*

1,2-dibromo-3-chloropropane
    (DBCP): 10, 15, 52, 54
6-Year Implementation Plan: 68
Advisory Committee on
    Nematology: 26-27
affiliation: 33, 39-40
agricultural research: 51, 69-70, 78
agricultural society: 1
American Institute of Biological
    Sciences (AIBS): 24-25, 28, 39
American Phytopathological Society
    (APS): 21, 28-29, 33, 39-42, 59
*Annals of Applied Nematology*: 63-64
applied research: 63
autonomy: 27, 41
bankruptcies: 56
basic research: 66
biological control: 53, 66, 68
biotechnology: 66, 68, 80-81, 84
bioterrorism: 81-82
Brazilian Nematological Society: 76
burrowing nematode: 10
Carter, Walter: 10
charter members: 28, 30, 58-59, 65
Civil War: 2
Clean Water Act: 53-54
Coalition on Funding Agricultural
    Research Missions (CoFARM): 78
Cobb, Nathan A.: 1, 4-5, 46-47
Committee of Four: 27-29
Committee of Nine: 29-33
constitution: 30, 33, 35-36, 39, 49,
    57-61, 79, 107-122, 123-126
cover design: 46-48
crop loss: 5, 7
crop rotation: 3, 17, 55
departments of nematology: 69-70
dichloropropene-dichloropropane
    mixture (D-D): 10-11, 54
Economic Nematology Endowment: 79

Egyptian Society for Agricultural
    Nematology: 76
Entomophilic Nematology
    Endowment: 79
Environmental Protection Agency
    (EPA): 53-54, 78
environmental quality: 67
ethylene dibromide (EDB): 10-11, 54
European Society of Nematologists
    (ESN): 24, 63, 76
experiment stations: 14, 20, 71
Federal Water Pollution Control Act
    Amendment: 53
First International Congress of
    Nematology (FICN): 61-63
FLAVRSAVR tomato: 67
fumigant nematicide: 10, 15, 66
fundamental science: 66
genetic engineering: 66-67
genomics: 66, 86
GI bill: 18
global cooperation: 75
golden nematode: 9, 24
ground water contamination: 54
Hatch Funds: 68-69
hidden enemy: 9
Homeland Security: 82
Industrial Revolution: 2
inflation: 70
information technology: 66
Integrated Pest Management (IPM):
    55
International Federation of
    Nematological Societies (IFNS):
    75-76
International Society for Plant
    Pathology: 61
Italian Society of Nematology: 76
Japanese Nematology Society: 76
John M. Webster Outstanding

## 134 Agriculture, Nematology and SON

Student Award: 79
joint meeting: 41
*Journal of Nematology*: 45-49
K. R. Barker-IFNS Endowment: 79
low-input sustainable agriculture (LISA): 56
logo: 46-49
membership dues: 39, 46
molecular biology: 52, 65-72, 84
molecular genetics: 66
Nathan A. Cobb Nematology Foundation: 79-81
natural resources: 72
nematicides: 10-12, 15, 29, 39, 41-42, 63, 66-67, 83
nematode management: 54, 64
*Nematologica*: 24
Nematological Society of Southern Africa: 76
Nematology Committee of APS: 26, 39, 41-42
Nematology Newsletter (NNL): 22, 27, 30, 32, 43, 47
North Central Regional Project: 13
Northeastern Regional Project: 13
operation manual: 59
Organization of Tropical American Nematologists (OTA): 63
Pakistan Society of Nematology: 76
*Plant and Soil Nematodes: Societal Impact and Focus for the Future*: 73-74
Presidents' Endowment: 79-80
questionnaire: 29-32, 44
Regional Research Project: 13
Research and Marketing Act: 13
Russian Society of Nematology: 76
S-19 Technical Committee:
 *see also* Southern Regional Project
satellite technology: 66, 83

science of nematology: 10, 27, 37-38, 47, 53, 57, 68, 72, 74, 81, 84-85
separate society: 25-29
Shell Nematology Workshop: 15, 24
short course: 14-15, 18
*Silent Spring*: 53
societal impact: 54
Southern Regional Project (S-19 Technical Committee): 14-22
Twenty-fifth Anniversary: 62-65
Silver Jubilee:
 *see also* Twenty-fifth Anniversary
soybean cyst nematode: 7, 9, 21
sugar beet cyst nematode: 3, 15, 24, 69
sustainability: 56
sustainable agriculture: 7, 38, 56
symposia: 23-24
Symposium: 24-25, 37, 44
technological revolution: 66
technology: 12, 51-52, 66-68, 72-73, 85
tobacco cyst nematode: 9
transgenic: 68
US Environmental Protection Agency (EPA): 53-54, 78
*Vistas on Nematology*: 62
water quality: 53
Western Regional Project: 13
workshops: 14-19, 24
yield losses: 8